Climate Adaptation Santiago

Kerstin Krellenberg • Bernd Hansjürgens
Editors

Climate Adaptation Santiago

Springer

Editors
Kerstin Krellenberg
Department of Urban
and Environmental Sociology
Helmholtz Centre
for Environmental Research - UFZ
Leipzig
Germany

Bernd Hansjürgens
Department of Economics
Helmholtz Centre for
Environmental Research - UFZ
Leipzig
Germany

ISBN 978-3-642-39102-6 ISBN 978-3-642-39103-3 (eBook)
DOI 10.1007/978-3-642-39103-3
Springer Heidelberg New York Dordrecht London

Library of Congress Control Number: 2014933698

Printed on acid-free paper

Springer is part of Springer Science+Business Media (www.springer.com)

Foreword

The effects of climate change in the Metropolitan Region of Santiago are a concern of the local authorities and there is genuine interest in tackling them adequately. The topic is new, however, and has only recently been considered in Chile. Since advances in climate change predictions in the last few years had not yet reached a regional scale, recommendations on regional adaptation to the adverse effects of such changes were non-existent. Consequently, the German cooperation proposal of launching a project to develop climate change projections and make recommendations for response on a regional scale in the Metropolitan Region of Santiago de Chile gained immediate regional attention and approval. The alliance between the German Helmholtz Centre for Environmental Research and scientific experts from the Pontificia Universidad Católica de Chile, the Universidad de Chile, and the Karlsruhe Institute for Technology was an added trust factor. The ClimateAdaptationSantiago (CAS) project, which was carried out in the Metropolitan Region of Santiago de Chile, is an example of how regional public policies on issues pertaining to climate change adaptation can be encouraged and supported on a scientific-technical basis.

It is important to highlight the way in which numerous regional public and private actors were consulted and participated in the process of disseminating and verifying the scientific advances made in climate change predictions and in the development of suitable adaptation measures. This democratic act lends more weight to the results of the CAS project and helps to generate action, which is the task of public institutions. Validation of public action by a broad range of actors from the private sector, NGOs, environmental organizations, and public institutions facilitates both its effectivity and its implementation.

It should likewise be emphasized that the CAS project stimulated the coordination and organization of public institutions around the topic of climate change. Parallel to project initiation, the Ministry of Environment began to develop its capacity to coordinate public institutions in the context of climate change issues, in accordance with a law passed in February 2010 assigning these competencies to the Ministry. The latter's participation in the CAS project made it possible to put them into practice. At the same time, the Regional Government, the first institution to embrace the project, created a working group within the Environmental Commission of the Regional Council, entitled the Sub-commission for Climate Change. It was composed of regional councillors and advised by a representative of the Regional Secretary of the Ministry of Environment, a representative of the Regional Government and two representatives of the CAS project. This can be understood as a significant political act, since it allowed for the creation of a permanent working group on the topic of climate change within the Regional Council, the entity that decides on public regional investment on behalf of the Regional Government. Under the aegis of the CAS project, the working group took the decision to follow up on what had been initiated by the project, to incorporate the newly acquired knowledge in regional decision-making, to disseminate the relevant information among the population throughout the region, and to monitor the outcome of the project - the evolution of the Regional Climate Change Adaptation Plan.

Santiago de Chile, August 2013

Rodrigo Robles (Regional Government)
Osvaldo Aravena (Member of the Regional Council)
Jaime Rovira (Ministry of Environment)

Preface

This book was inspired by the observation that climate change and urbanization are two ongoing and interwoven processes. Given this strong interlinkage, cities around the world have begun to design regional and local climate change strategies with mitigation and adaptation elements to address the adverse effects of climate change. Megacities and large agglomerations play a significant role here as they tend to be key emitters of greenhouse gases and are heavily affected by the consequences of climate change. Many of these same cities have developed innovative strategies and transition options in response to climate change impacts. It is the complexity and diversity of ongoing processes and governance structures, as well as of social and economic heterogeneity that makes these cities an exceptionally challenging research object. A reasonable response to climate change that is also feasible calls for multi-dimensional, multi-level and multi-scale approaches leading to a greater understanding of the complexity of such processes.

This is the fundamental principle of the book. It describes the integrative inter- and transdisciplinary (IIT) approach used in the development of a Regional Climate Change Adaptation Plan for the Metropolitan Region of Santiago de Chile, including concrete adaptation measures and the evaluation and implementation of such a plan at the science-policy interface. The topics addressed are regional climate change, climate change impacts, adaptation needs and measures, and the implementation of these measures at the urban-regional level. The entire "chain" of analysis is exemplified for one case city. It builds on scientific analyses undertaken during a process initiated by social and natural scientists—an intensive participatory process that embraced a wide range of actors and stakeholders from the public and private sectors, civil society and academia, and a mutual learning network across megacities in Latin America.

The book draws on the international research project ClimateAdaptation-Santiago (CAS) (http://www.ufz.de/climate-adaptation-santiago), a combined enterprise of two German research institutes of the Helmholtz Association (Karlsruhe Institute of Technology (KIT) and Helmholtz Centre for Environmental Research—UFZ), three partner organizations in Latin America (Universidad de Chile, Pontificia Universidad Católica de Chile, Economic Commission for Latin America and the Caribbean of the United Nations—ECLAC/CEPAL), and the two main climate change decision-making entities in Santiago de Chile (Regional Government of the Metropolitan Region of Santiago de Chile and Regional Secretary of the Ministry of the Environment). The project involved about twenty-five researchers in Germany and Chile and sixty local stakeholders representing approximately forty different Chilean organizations. Researchers and decision-makers from the cities of Bogotá, Buenos Aires, Lima, Mexico and Sao Paulo also contributed to this inter- and transdisciplinary project, which was carried out between 2009 and 2013.

This book presents the overall results of the enterprise. Although an edited volume, it differs considerably from a collection of papers, since the chapters follow an overarching structure and analysis. Each of the 12 chapters builds on the previous one, providing a coherent picture of climate change impacts and adaptation analysis in urban areas. It develops transferable solution and acts as an incentive to other cities that still face the challenge of designing comprehensive climate change response strategies.

Acknowledgements

We would like to thank all those who made the commitment to contribute to the success of the ClimateAdaptationSantiago (CAS) project. Our thanks go first of all to the more than 25 contributing authors of this book.

In terms of the overall process of the CAS project we are particularly grateful to Osvaldo Avarena (Consejero of the Regional Government of the Metropolitan Region of Santiago de Chile), Rodrigo Robles (Regional Government of the Metropolitan Region of Santiago de Chile), and Jaime Roviera (Ministry of Environment), who were instrumental in pushing the process forward on the political agenda in Santiago de Chile.

We would furthermore like to thank the representatives from the public sector, private enterprises, academia and civil society organizations who contributed greatly to the success of the project by taking an active part in the participatory process (the ten roundtable meetings organized in Santiago de Chile). In this context, our special thanks go to Jonathan Barton and Jordan Harris for making their own contacts available and pushing the roundtable process forward locally. In addition, we are immensely grateful to the universities concerned and the Regional Government for hosting the roundtable meetings.

We also thank the participants of the three workshops organized in the frame of the Regional Learning Network for their interest and collaboration. They came from six Latin American megacities to share their experience of climate change adaptation with us. In this regard, special thanks go to the UN CEPAL, who hosted the workshops, not least Ricardo Jordán, Johannes Rehner, Benjamín Infante and Alejandra Pérez.

Sunniva Greve did an incredible job as language editor of the book and we would like to express our special thanks for her patience throughout the process.

The overall production of the book over a period of almost two years would not have been possible without the continuous and outstanding efforts of our colleague Katrin Barth, who tirelessly supported us with editorial work and helpful suggestions. Her contribution was key to developing a coherent manuscript.

Financial support was provided by the German Federal Ministry for the Environment, Nature Conservation and Nuclear Safety (BMU) through its International Climate Initiative.

Leipzig, October 2013

Kerstin Krellenberg
Bernd Hansjürgens

Contents

List of Figures

List of Tables

Part I

Challenges for Urban Climate Change Adaptation

1 Introduction

Kerstin Krellenberg and Bernd Hansjürgens

Abstract

The introductory chapter focuses on the principal topics and conceptual frameworks to be addressed in this book. Placing urban climate change adaptation centre stage, it provides a general perspective on the relation between climate change and urbanization, and on adaptation and mitigation, highlighting the specific challenges for Latin America and Santiago de Chile in this context. Furthermore it defines the aims of the book and its research approach, and clarifies its essential contribution to current debates on urban climate change adaptation. In conclusion it presents an overview of the subsequent chapters.

Keywords

Megacities • Climate change adaptation • Adaptive capacity • Santiago de Chile

1.1 Urban Climate Change and Adaptation Needs

There is a growing consensus today that warming of the climate system is unequivocal and that most of the observed increases in global average temperatures are very likely due to human activity (IPCC 2007, Annex B, Glossary of Terms). The effects of climatological alterations in temperature and precipitation have already been recognized. Uncertainties about the extent of climate change and the magnitude of its impacts at regional and local level, however, still exist. Consequently uncertainty in modelling climate changes and the related impacts is high.

K. Krellenberg (✉)
Department of Urban and Environmental Sociology, Helmholtz Centre for Environmental Research – UFZ, Permoser Str. 15, 04318 Leipzig, Germany
e-mail: kerstin.krellenberg@ufz.de

B. Hansjürgens
Department of Economics, Helmholtz Centre for Environmental Research – UFZ, Permoser Str. 15, 04318 Leipzig, Germany

K. Krellenberg and B. Hansjürgens (eds.), *Climate Adaptation Santiago*,
DOI 10.1007/978-3-642-39103-3_1,

Interwoven with climate change is another outstanding process: urbanization. For the first time in history and now an upward trend, more than 50 % of the world's population lives in cities. Its distinction as a global phenomenon notwithstanding, urbanization does not take place homogeneously across continents and countries or even within the borders of a single country. Most future growth is expected to occur in medium-sized cities in the developing world (United Nations 2008).

Latin America and the Caribbean, the focus of this book, show an urban population of 77.5 % (United Nations 2008). The countries with the highest percentage of urban population in the region are Chile, Argentina, Uruguay and Venezuela, where up to the latter half of the twentieth century the population was concentrated in a small number of very large cities (Jordán et al. 2010). At the beginning of the twenty-first century, these population density rates gradually decelerated and led to more diversified patterns of urban concentration in the direction of smaller cities (Rodriguez 2008). The region's population nevertheless continues to converge in its principal cities; as economic, political, and cultural centres, they shoulder a heavy concentration of knowledge, capital and human resources. In the year 2000, approximately 20 % of the total population of Latin America lived in cities of more than five million inhabitants, surpassing all other world regions. These populous concentrations point towards another trend: the appearance in the urban agglomeration landscape of megacities with at least ten million inhabitants (United Nations 2008) perhaps the most visible expression of rapid urbanization.

Cities can be climate change culprits and victims simultaneously. They show specific climates, consume much of the world's energy and produce most of its greenhouse gas emissions. As a result of their concentration of people and services, their centralizing of economic power, their infrastructure and their high demand for natural resources (e.g., water, energy, food), cities are also heavily affected by the impact of climate change. At the same time they are in possession of numerous benefits and opportunities for transition, enabling them to address and respond to this impact.

Considerable differences exist, however, between cities of the "global north" and those of the "global south". Whereas the majority of the former are key emitters of greenhouse gas, the latter have already begun to feel the consequences of global climate change in the form of increased risk exposure (to sea level rise, landslides, floods, droughts, heat waves), since population growth tends to go hand in hand with the expansion of urban areas in high risk environments. A number of these trends are exacerbated by high socio-spatial differentiation leading to frequent placing of "the poor" in hazard-prone areas (Sherbinin et al. 2007; Kuhlicke et al. 2012), where living conditions are fraught with danger as a result of using building materials that are inappropriate in relation to coping with hazards. Furthermore, the governance structures of these cities are complex and can therefore amplify or even produce risks to and negative impacts on human security.

The appearance of climate change on the international political agenda produced responses at different levels of governance: global, international, national, regional and local. Given that a specific level of climate change impact is irreversible and the rate of change very slow, CO_2 concentrations are not expected to decrease significantly, even if the world were to suddenly shift to a net zero carbon economy (IPCC 2007).

Hence adapting to the consequences of climate change is vital and in conjunction with mitigation measures a key component of climate policy. Cities around the world have begun to react. Many have become 'proactive' initiators of climate strategies, at times moving ahead of national and global agenda-setting, preparing for the risks and opportunities related to climate change, and bundling long-term development policy-making with the climate agenda, including elements of mitigation and adaptation (Berrang-Ford et al. 2010; Heinrichs et al. 2013). Since the beginning of the 1990s, local governments and other urban actors have taken the initiative to adjust their structures, practices and processes in response to changing climate conditions and their adverse effects. Bulkeley (2010) distinguishes two phases of urban response to climate change: a first phase pioneered by local governments predominantly in the global north, which saw the launching of concrete local policy initiatives to reduce the consumption of environmental resources, particularly energy. The second phase, which got under way in the early 2000s, was more political in nature. It embraced a wider array of climate policy issues and has become increasingly sensitive to such concerns as risk and vulnerability, and how to adapt to these factors.

Multiple interlinking processes of climate change now affect several sectors of society simultaneously and, coupled with the complexity of the governance structures involved, call for multi-dimensional, multi-level and multi-scale approaches if climate change in the urban areas is to be met with an adequate response (Seto et al. 2010; Heinrichs et al. 2013).

Adaptation to climate change has been defined by different disciplines in numerous ways (e.g., Smit et al. 2000). The IPCC (2001) describes it as the adjustment of structures, practices and processes to changing climate conditions and their impacts. Klein et al. (2007) state that dealing with adaptation requires an understanding of the vulnerability of societies and ecosystems to climate change impacts, their capacity to respond and the socio-economic costs it entails. The term adaptation used in this book describes the reaction to risks and vulnerabilities of selected key sectors as they occur under climate change. It refers to the adjustment of laws, programmes, plans and measures in order to curb negative climate change impacts at urban-regional and municipal level. In this context, tackling the complexity of climate change and the interweaving processes involved requires cross-cutting action between sectors (e.g., Shaw et al. 2007). Climate action plans and adaptation strategies are some of the activities with the potential to unite sectors and levels of decision-making within the scope of an integrated planning approach (Healey 1992).

While the need for adaptation and the idea of adaptive response action has been widely acknowledged, obstacles to and constraints on adaptation plans, strategies and measures under real-world conditions still exist and hamper their successful implementation. It is at this point that the current book seeks to go a step further: it presents a conceptual framework (cf. Sect. 1.4) that allows climate change adaptation measures to be developed, prioritized, selected and implemented, and subsequently embedded in a Regional Climate Change Adaptation Plan, taking the Metropolitan Region of Santiago de Chile (MRS) as a case study (Krellenberg 2012).

The remaining sections of this introductory chapter stress the role of adaptive capacity as a decisive factor in responding to global climate change (Sect. 1.2). The case study area is introduced in Sect. 1.3, while the objectives and research approach are laid down in Sect. 1.4. The concluding Sect. 1.5 provides an overview of the subsequent chapters.

1.2 Response Action to Climate Change: Adaptive Capacity as a Decisive Factor

The advance of climate change on the international political agenda has by the same token produced an increase in the number of analytical studies on the topic. Several studies focus on 'adaptive capacity', explaining how local governments address the issue of climate change and identifying successes and failures. Here adaptive capacity describes the potential of a system, region or community to adjust to the adverse effects of climate change including extremes (Smit and Wandel 2006). It defines the ability to prepare for climate change risks and opportunities (proactive or autonomous adaptation) and to cope with or adjust to potential and traceable negative impacts (reactive adaptation). Adaptive capacity is thus closely associated with the concept of coping capacity. The latter describes the ability to deal directly with extreme events on a short time horizon, while the time frame of 'adaptive capacity' is typically longer and builds, for example, on learning processes (Yohe 2001). Gupta et al. (2010:461) interpret adaptive capacity as the "inherent characteristics of institutions that empower social actors to respond to short- and long-term impacts either through planned measures or through allowing and encouraging creative responses from society both ex ante and ex post". This perspective includes an assessment of the extent to which these features enable society to cope with climate change or encourage it to alter such institutions to achieve this purpose.

A number of conceptual approaches interpret the capacity to respond, cope or adapt as a dimension or component of vulnerability (Heinrichs et al. 2013). In one of the earliest definitions of vulnerability, Chambers (1989) emphasizes that 'capacity' is merely one element of the dual nature of vulnerability and links it to exposure to the stress of climate events. Others like Bollin and Hidajat (2006), for example, prefer to interpret vulnerability and capacity as separate entities. This perspective is widely accepted in disaster risk research and management practice. It sees capacity as a more independent phenomenon, defining it as the general ability to confront disasters and reduce risks (e.g., UN/ISDR 2006).

As the recent literature on climate change indicates, elements that drive local action—a possible explanation of its potential success—fall broadly into three categories of local adaptive capacity: ability, willingness and the enabling/disabling context (Yohe 2001; Zahran et al. 2008; Burch and Robinson 2007; Tompkins and Adger 2005). Following the IPCC (2007, Annex B, Glossary of Terms), adaptation can occur in different ways and on different levels:

- Firstly, adaptation can be *anticipatory* or proactive, i.e., "adaptation that takes place before impacts of climate change are observed" by preparing for climate change risks and opportunities, or *reactive adaptation*, which is "adaptation that takes place after impacts of climate change have been observed".
- Secondly, *private adaptation* "is initiated and implemented by individuals, households or private companies", and is "usually in the actor's rational self-interest", whereas *public adaptation* "is initiated and implemented by governments at all levels", and is "usually directed at collective needs".
- Finally, there is *autonomous or spontaneous adaptation*, which "does not constitute a conscious response to climatic stimuli but is triggered by ecological changes in natural systems and by market or welfare changes in human systems", while *planned adaptation* "is the result of a deliberate policy decision, based on an awareness that conditions have changed or are about to change and that action is required to return to, maintain or achieve a desired state".

The focus of this book lies on the adaptive capacity of urban decision-makers and their ability to act. Here adaptation is based on a science-initiated process (cf. Sect. 1.4 and Chap. 9) and can be regarded as public, planned adaptation. Despite its consideration of current and future climate change effects, this form of adaptation is anticipatory rather than reactive in nature, as current impacts in the study region Santiago de Chile are relatively low (see Chaps. 4–8). The approach takes into account that responses to climate change involve multiple actors and decisions. Advancing the governance of climate change across all government levels and stakeholders is therefore crucial if policy gaps between national policy frameworks and local action plans (vertical integration) are to be avoided and cross-scale learning between the relevant local and regional government departments and institutions (horizontal integration) encouraged (Corfee-Morlot et al. 2009).

In this sense, a multi-level governance approach to climate change response should take into consideration that national governments cannot implement their climate strategies effectively without close collaboration with regional and local governments as agents of change. Hence learning, information transmission and cooperation between cities or regions and national governments (see Chap. 10), and the refinement of coordination across national ministries implementing cross-sectoral programmes, as many climate change policies require, is vital (Corfee-Morlot et al. 2009). All of these aspects have been incorporated in a detailed analysis of anticipatory, public, planned adaptation at urban-regional level for the Metropolitan Region of Santiago de Chile. The case introduced in the following section gives a deeper insight into the study region.

1.3 The Case Study Region: The Metropolitan Region of Santiago de Chile

Today approximately 85 % of the Chilean population lives in urban centres; close to a third is concentrated in the capital city. The population of the Metropolitan Region of Santiago de Chile (MRS) is estimated at six million inhabitants (INE

2002) and shows an average annual increase of a 1.7 % (1995–2000) (CEPAL 2000). Urban expansion in the MRS is mainly directed towards the outskirts, and land conversion for urban expansion is taking place at a higher rate than ever before (Borsdorf and Hidalgo 2007). Expansion beyond the political-administrative boundaries of the MRS exceeds the capacity of existing administrations and their organizational structures to cope satisfactorily with the dynamics of this development.

Although Santiago de Chile is not seen as a megacity (if population figures of ten million are taken as a defining characteristic), it is a typical example of Latin America's largest cities, all of which face wide-ranging problems representative of others in the region (Jordan et al. 2012). As a result of its advanced urbanization growth patterns and the attendant demographic transformation, Santiago de Chile is commonly referred to when it comes to illustrating the stages of urban growth in Latin American megacities. It is also an excellent example of the pressures and consequences that urbanization entails, and the corresponding political responses. In comparison with other cities in the region, Santiago de Chile shows the lowest level of poverty but a particularly high level of socio-spatial segregation (Jordán et al. 2010). Private actors are assigned a vital role, for example, as providers of water services and housing.

Even without the challenge of climate change, the MRS suffers from administrative fragmentation, asymmetries of power, and overlapping democratic constellations (Chuaqui and Valdivieso 2004; Orellana 2009), consisting as it does of six provinces with a total of 52 independent municipalities, each of which is tasked with a local government (Nuissl et al. 2012). The MRS is administrated by the regional government (*Gobierno Regional Metropolitano de Santiago*). Its highest regional representative (the *Intendente*) is appointed by the president of Chile, a regional council (regional councillors are elected by municipal councillors), and the regional ministerial secretariats (*SEREMI*) (Chuaqui and Valdivieso 2004). There is an obvious absence of overall guiding instruments to shape sectoral and territorial policies, plans and investments, and to monitor their performance over time. In addition, governments prefer short-term action as opposed to long-term planning (Barton and Kopfmüller 2012). The obstacles to more integrated planning within the confines of existing structures are rooted in the sector divide and the short-term planning agenda. This is particularly challenging when it comes to designing effective local, long-term climate adaptation measures under consideration of, for example, existing measures and plans.

In sum, the current situation calls for greater multi-level planning, sector policy integration, and sustainable strategic and long-term action. It furthermore warrants the readjustment and redistribution of national and local responsibilities, coupled with new structures, strategies and instruments to pursue and implement policies.

Climate change response in Chile is guided by the United Nations Framework Convention on Climate Change (UNFCCC) and the national Climate Change Strategy developed in 2006. The National Climate Action Plan (*Plan de Acción Nacional de Cambio Climático*) elaborated by *Comisión Nacional del Medio Ambiente* (CONAMA 2008) addresses both mitigation and adaptation issues, but focuses primarily on mitigation aspects and *Clean Development Mechanism* (CDM) activities. Exposed (eco)systems, sectors and vital infrastructure, and the associated economic effects are taken into consideration. The underlying scenario-based vulnerability assessment of (natural) systems can be described as an 'outcome vulnerability' approach (O'Brien et al. 2007). It takes into account the adverse effects of climate change that heighten the exposure 'probability' of systems and sectors to dangerous situations. It marks vulnerability as the 'end point' of a sequence of analyses and pursues a top-down perspective. The IPCC definition of vulnerability with regard to the effects of global climate change is characteristic of this approach (IPCC 2007:883). Adaptive governance responses, following the 'outcome vulnerability' interpretation, often fail to consider social groups and urban infrastructure, both of which are exposed to climate change impacts (Heinrichs et al. 2011). Consequently the National Climate Action Plan has no explicit urban focus, apart from the coastal cities (Heinrichs and Krellenberg 2011; Krellenberg and Heinrichs 2010) that frame the elaboration of adaptive measures at urban-regional level in the MRS. Hence in terms of spatial planning, the instruments currently at the disposal of the MRS have not yet considered climate change specifically.

In 2010, the former CONAMA, which published the National Climate Action Plan, was transformed into the Ministry of the Environment. The Ministry now has the overall responsibility for climate change issues and is charged with establishing sectoral adaptation plans as well as a national adaptation plan. Key institutions for the control of instruments mentioned in the National Climate Action Plan at regional level and among those that implement climate adaptation strategies at urban-regional level are the Regional Government (GORE) and the Regional Secretariat of the Ministry of the Environment (SEREMI Medio Ambiente) (Barton 2009). This makes both entities key stakeholders in the process of developing climate change adaptation measures (cf. Chap. 9).

1.4 Objective and Research Approach of the Book

In view of the aspects mentioned above, not least the lack of climate change response action at urban-regional level in Santiago de Chile, the overall **objective** of this book is to demonstrate the development, prioritization and contextualization of concrete measures for a Regional Climate Change Adaptation Plan in an integrative inter- and transdisciplinary (IIT) approach. Taking the Metropolitan Region of Santiago de Chile as a case study, it highlights the most pressing issues of land-use change, its linkage to natural hazards, the associated vulnerabilities, and the supply of and demand for water and energy. It considers regional climate changes and their impacts, and the development of adaptation measures with the ultimate aim of producing a Plan to be accepted and implemented by political authorities and the public alike.

To reach this overall objective, three overarching **research questions** are addressed:

- What are the risks and vulnerabilities associated with regional climate change in the MRS?
- What measures and strategies constitute an adequate response to the challenges of climate change?
- How can measures developed on a participatory basis be transformed into a decision-making response?

  A fourth question arises on the meta-level:
- How can (and should) a science-cum-multi-stakeholder participatory approach to climate change adaptation be organized and implemented?

  Focusing on these research questions, the book contains three distinguishing features representative of its overall **research approach** (Fig. 1.1):
- The first characteristic feature of this book is its consideration of the entire chain of analyses from changes in climate to the related impacts and vulnerabilities for the case study region, the Metropolitan Region of Santiago de Chile:
  - In a first step, an assessment of regional climate changes is undertaken. It builds on downscaled information on major changes to the global climate at urban-regional level in Santiago de Chile. The spatial presentation of this information is new and contributes greatly to the existing knowledge on climate change in the MRS (cf. Chap. 2).
  - Based on these estimates, current impacts and those expected for the year 2050 in the fields of energy, water and land use and the attendant vulnerabilities are assessed (Chaps. 4–8), applying the explorative scenario approach adopted by Kopfmüller et al. (2009, 425 ff.). The latter provides the framework for potential socio-economic changes (cf. Chap. 3).
  - A set of adaptation measures based on the identified need for action is elaborated as part of a Regional Climate Change Adaptation Plan for the MRS. It includes the identification, prioritization and selection of specific measures, the analysis of their potential impacts and interdependencies, and the implementation hurdles to be overcome.
- The second distinguishing feature of the book is its interdisciplinary approach. It touches on numerous scientific disciplines ranging from the natural (e.g., meteorology, hydrology, climatology) to the social science disciplines (e.g., planning, economics, law, political science, urban sociology) and calls for interdisciplinary investigations. It is precisely the complexity and interweaving of the processes of climate change that makes the interdisciplinary approach presented in this book indispensable.
- The third feature of distinction—closely related to the second—is the perception that a science-policy approach is crucial to the successful elaboration of climate change adaptation strategies and their implementation. The multiple scales of political decision-makers and stakeholders involved render climate change policies exceptionally complex, bringing the science-policy interface increasingly to the fore in contemporary climate change research on response capacity generation. From the outset, stakeholder involvement has been pivotal to the exchange of 'usable' information with decision-makers within the participatory processes of developing adaptation measures, presented as a transdisciplinary

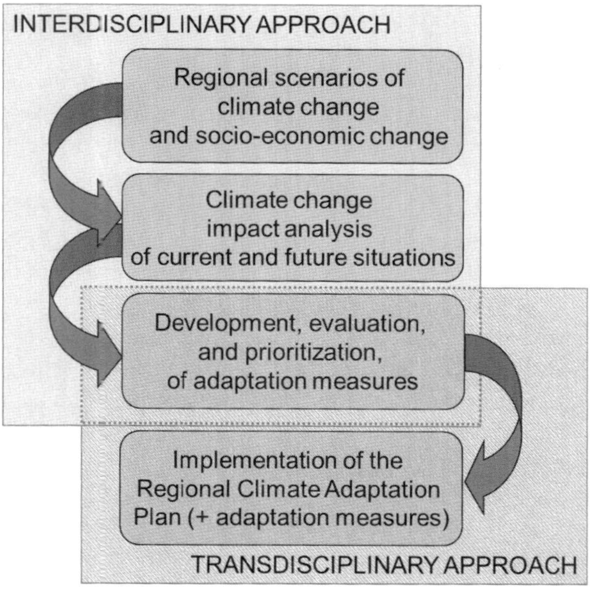

Fig. 1.1 Integrative inter- and transdisciplinary research approach (*Source*: Authors' own presentation)

approach in this book (Chap. 9). This holds true for at least two reasons: (1) It guarantees to a large extent that prevailing political and social demands and inquiries are taken into account, and (2) it makes adaptation needs visible and generates adequate responses, taking social, cultural and institutional constraints based on adaptive capacities into account (e.g., Adger et al. 2005; Corfee-Morlot et al. 2011; Smith and Stern 2011).

This last aspect is also emphasized by Adger (2003) when he states that adaptation strategies to climate change are equally dependent on the ability of individuals and communities to act collectively, and involve intervention and planning by the state. The importance of collective action and social capital-building has been studied as a vital component of institutional processes to create climate change adaptation capacity.

Based on these theoretical-conceptual reflections, the core element of the participatory process in Santiago de Chile was the organization of ten roundtable meetings over a two-year period with key institutional representatives from a wide array of sectors and levels of society (cf. Chap. 9).

1.5 Overview of the Book

The chapters in this book follow the overarching structure and analysis presented in Sect. 1.4, spanning climate change analysis on a regional scale, through the analysis of impacts and vulnerabilities to the development of adaptation measures in key sectors and the discussion of implementation issues. Thus the individual parts and chapters build on the foregoing and provide a coherent picture of the entire climate adaptation process in an urban area.

Part II "Climate change impacts on the urban-regional level of Santiago de Chile" consists of two chapters that address the methodology of downscaling global climate changes to the regional level in order to demonstrate their impact at the urban-regional level of Santiago de Chile and show how working with explorative scenarios techniques allows for the assessment of these changes.

- Chapter 2 on "Downscaling climate changes for Santiago: what effects can be expected" presents the main results on climate conditions, historic trends and the expected climate changes at urban-regional level in Santiago de Chile. It includes downscaling climate models to the regional level and a discussion on uncertainties and the added value of the information generated. This information serves as the background to the analysis of climate change impacts in Part III.
- Chapter 3, "Scenarios for future developments" presents a future perspective on climate change impacts by combining the results on future climate change with socio-economic developments up to 2030 and 2050. It presents and justifies the explorative scenario approach applied in Part III, Chaps. 4–8.

Part III "Climate change impacts on the urban-regional level of Santiago de Chile: key sectors and vulnerabilities" comprises five chapters, four of which relate to a specific sector, while the final chapter describes the interlinkages between them. The result of in-depth empirical investigation, these chapters seek to estimate the impact and consequences of climate change in specific climate-sensitive sectors. In line with the priorities of the Chilean National Action Plan and previous studies on Santiago de Chile, the fields of water, energy and land-use change were identified as particularly relevant in the context of climate change impacts. Current and future conditions are presented and first steps discussed to confront the challenges identified in these fields.

- Chapter 4 focuses on the water sector, which is heavily affected by climate change.
- Chapter 5 looks at the energy sector, one that greatly depends on water and other energy imports. It is expected to gain in significance with the prediction of higher temperatures and less rainfall.
- Chapter 6 deals with the relationship between changing land use/cover, climate change, and flood and heat hazard under the pressure of climate change.
- Chapter 7 adds socio-economic aspects that are highly relevant when addressing the vulnerability of people and housing. It assesses hazard-related exposure (flood and heat) as analysed in Chap. 6.
- Chapter 8 gives a brief synthesis of the findings, bridging the sector divide with a discussion on interlinkages between the challenges identified in the previous chapters.

Part IV on "Adaptation Strategy: Developing measures and implementation" turns to the question of how to respond to the urgent climate change impacts in an urban context such as Santiago de Chile. It contains four chapters:

- Chapter 9 concentrates on the transdisciplinary approach to the development of adaptation measures. It describes the development and establishment of a combined multi-stakeholder participatory approach.

- Chapter 10 discusses opportunities and obstacles to the implementation of the adaptation measures identified. It includes a description of capacities and competences as decisive factors for the city's response to climate change impacts.
- Chapter 11 represents the main findings of a Regional Learning Network established by scientists and decision-makers from six Latin American megacities (Buenos Aires, Bogotá, Lima, México, Santiago de Chile and Sao Paulo). It summarizes the most important outcomes of the Network, the aim of which was to strengthen mutual learning by scientists and decision-makers in the Latin American region in the context of urban climate change adaptation. It discusses the challenges that the transfer of approaches, methods and measures entails.
- Chapter 12 draws several conclusions and discusses lessons learned. It presents, describes and evaluates the overall results of the book, focusing on the Regional Climate Change Adaptation Plan and the discussion of appropriate implementation mechanisms.

References

Adger, W. N. (2003). Social capital, collective action and adaptation to climate change. *Economic Geography, 79*, 387–404.

Adger, W. N., Arnell, N. W., & Tompkins, E. L. (2005). Successful adaptation to climate change across scales. *Global Environmental Change, 15*, 77–86.

Barton, J. R. (2009). Adaptación al cambio climático en la planificación de ciudades-regiones. *Revista de Geografía Norte Grande, 43*, 5–30.

Barton, J. R., & Kopfmüller, J. (2012). Sustainable urban development in Santiago de Chile: Background - Concepts - Challenges. In D. Heinrichs, K. Krellenberg, B. Hansjürgens, & F. Martínez (Eds.), *Risk Habitat Megacity* (pp. 65–86). Heidelberg: Springer.

Berrang-Ford, L., Ford, J. D., & Paterson, J. (2010). Are we adapting to climate change? *Global Environmental Change, 21*, 25–33.

Bollin, C., & Hidajat, R. (2006). Community-based disaster risk index: Pilot implementation in Indonesia. In J. Birkmann (Ed.), *Measuring vulnerability to natural hazards: Towards disaster resilient societies* (pp. 271–289). New York: United Nations University Press.

Borsdorf, A., & Hidalgo, R. (2007). New dimensions of social exclusion in Latin America: From gated communities to gated cities, the case of Santiago de Chile. *Land Use Policy, 25*, 153–160.

Bulkeley, H. (2010). Climate policy and governance: an editorial essay. *Wiley Interdisciplinary Reviews: Climate Change, 1*, 1311–1313.

Burch, S., & Robinson, J. (2007). A framework for explaining the links between capacity and action in response to climate change. *Climate Policy, 7*(4), 304–316.

CEPAL. (2000). *From rapid urbanization to the consolidation of human settlements in Latin America and the Caribbean: A territorial perspective*. Santiago de Chile.

Chambers, R. (Ed.). (1989). *Vulnerability. How the poor cope*. Brighton: IDS.

Chuaqui, T., & Valdivieso, F. (2004). Una ciudad en busca de un gobierno: una propuesta para Santiago. *Revista de Ciencia Política, 24*(1), 104–127.

CONAMA. (2008). *Plan de Acción Nacional de Cambio Climático (2008-2012)*. Santiago: CONAMA.

Corfee-Morlot, J., Kamal-Chaoui, L., Donovan, M. G., Cochran, I., Robert, A., & Teasdale, P. J. (2009). *Cities, climate change and multilevel governance*. OECD environmental working papers 14.

Corfee-Morlot, J., Cochran, I., Hallegatte, S., & Teasdale, P. J. (2011). Multilevel risk governance and urban adaptation policy. *Climatic Change, 104*, 169–197.

Gupta, J., Termeer, C., Klostermann, J., Meijerink, S., Van den Brink, M., Jong, P., et al. (2010). The adaptive capacity wheel: A method to assess the inherent characteristics of institutions to enable the adaptive capacity of society. *Environmental Science and Policy, 13*, 459–471.

Healey, P. (1992). Planning through debate: The communicative turn in planning theory. *The Town Planning Review, 63*(2), 143–162.

Heinrichs, D., & Krellenberg, K. (2011). Climate adaptation strategies: Evidence from Latin American city-regions. In K. Otto-Zimmermann (Ed.), *Resilient cities: Cities and adaptation to climate change - proceedings of the global forum 2010 (local sustainability)* (Vol. 1, pp. 223–231). Dordrecht: Springer.

Heinrichs, D., Aggarwal, R., Barton, J., Bharucha, E., Butsch, C., Fragkias, M., et al. (2011). Adapting cities to climate change: Opportunities and constraints. In D. Hoornweg, M. Freire, M. J. Lee, P. Bhada-Tata, & B. Yuen (Eds.), *Cities and climate change: Responding to an urgent agenda* (pp. 193–224). Washington, DC: The World Bank.

Heinrichs, D., Krellenberg, K., & Fragkias, M. (2013). Urban responses to climate change: Theories and governance practice in cities of the Global South. *International Journal of Urban and Regional Research, 37*, 1865–1878.

INE (Instituto Nacional de Estadísticas). (2002). *Censo de Población y Vivienda*. Santiago: INE.

IPCC. (2001). *Climate change 2001. Impacts, adaptation and vulnerability*. Contribution of Working Group II to the third assessment report of the intergovernmental panel on climate change (IPCC). Cambridge: Cambridge University Press.

IPCC. (2007). *Climate change 2007. Impacts, adaptation and vulnerability*. Working Group II contribution to the intergovernmental panel on climate change fourth assessment report. Cambridge: Cambridge University Press.

Jordán, R., Rehner, J., & Samaniego, J. (2010). *Regional Panorama Latin America: Megacities and sustainability*. Santiago de Chile: United Nations.

Jordan, R., Rehner, J., & Samaniego, J. (2012). Megacities in Latin America: Role and challenges. In D. Heinrichs, K. Krellenberg, B. Hansjürgens, & F. Martínez (Eds.), *Risk Habitat Megacity* (pp. 19–35). Heidelberg: Springer.

Klein, R., Schipper, E.L.F., & Dessai, S. (2007). *Integration of mitigation and adaptation into climate and development policy: Three research questions*. Tyndall centre for climate change research working paper 40.

Kopfmüller, J., Lehn, H., Nuissl, H., Krellenberg, K., & Heinrichs, D. (2009). Sustainable development of megacities: An integrative research approach for the case of Santiago Metropolitan Region. *Die Erde, 140*(4), 417–448.

Krellenberg, K. (2012). A science-policy approach towards local adaptation planning: The case of Santiago de Chile. In K. Otto-Zimmermann (Ed.), *Resilient cities 2: Cities and adaptation to climate change – Proceedings of the global forum 2011 (local sustainability)* (pp. 233–240). Dordrecht: Springer.

Krellenberg, K., & Heinrichs, D. (2010). Urban Latin America under climate change: Do adaptation strategies of city-regions respond to the challenges? In W. Endlicher & F.-W. Gerstengarbe (Eds.), *Continents under climate change* (Nova Acta Leopoldina 384th ed., Vol. 112, pp. 267–274). Stuttgart: Wissenschaftliche Verlagsgesellschaft.

Kuhlicke, C., Kabisch, S., Krellenberg, K., & Steinführer, A. (2012). Urban vulnerability under conditions of global environmental change. Conceptual reflections and empirical examples from growing and shrinking cities. In S. Kabisch, A. Kunath, P. Schweizer-Ries, & A. Steinführer (Eds.), *Vulnerability, risk and complexity: Impacts of global change on human habitats* (pp. 27–38). Cambridge, MA: Hogrefe.

Nuissl, H., Höhnke, C., Lukas, M., Durán, G., & Rodriguez Seeger, C. (2012). Megacity governance: Concepts and challenges. In D. Heinrichs, K. Krellenberg, B. Hansjürgens, & F. Martínez (Eds.), *Risk Habitat Megacity* (pp. 87–108). Heidelberg: Springer.

O'Brien, K., Eriksen, S., Nygaard, L., & Schjolden, A. (2007). Why different interpretations of vulnerability matter in climate change discourses. *Climate Policy, 7*, 73–88.

Orellana, A. (2009). La gobernabilidad metropolitana de Santiago: la dispar relación de poder de los municipios. *EURE (Santiago), 35*(104), 101–120.

Rodriguez, J. (2008). Spatial distribution of the population, internal migration and development in Latin America and the Caribbean. In *Proceedings of the United Nations Expert Group meeting on population distribution, urbanization, internal migration and development*. New York, pp. 1–30.

Seto, K. C., Sanchez-Rodriguez, R., & Fragkias, M. (2010). The new geography of contemporary urbanization and the environment. *Annual Reviews of Environment and Resources, 35*, 167–194.

Shaw, R., Colley, M., & Connell, R. (2007). *Climate change adaptation by design: A guide for sustainable communities*. London: TCPA.

Sherbinin, A., de Schiller, A., & Pulsipher, A. (2007). The vulnerability of global cities to climate hazards. *Environment and Urbanization, 19*(1), 39–64.

Smit, B., & Wandel, J. (2006). Adaptation, adaptive capacity and vulnerability. *Global Environmental Change, 16*, 282–292.

Smit, B., Burton, I., Klein, R., & Wandel, J. (2000). An anatomy of adaptation to climate change and variability. *Climate Change, 45*, 223–251.

Smith, L., & Stern, N. (2011). Uncertainty in science and its role in climate policy. *Philosophical Transactions of the Royal Society A: Mathematical, Physical and Engineering Sciences, 369* (1956), 4818–4841.

Tompkins, E., & Adger, N. (2005). Defining response capacity to enhance climate change policy. *Environmental Science and Policy, 8*(6), 562–571.

UN Nations (Population Division). (2008). *World urbanization prospects. The 2007 revision*. New York: UN Nations.

UN/ISDR. (2006). *Hyogo framework for action 2005–2015*. Geneva: UN/ISDR. http://www.unisdr.org. Accessed 18 Feb 2010.

Yohe, G. (2001). Mitigative capacity: The mirror image of adaptive capacity on the emission side. *Climate Change, 49*(3), 247–262.

Zahran, S., Brody, S. D., Vedlitz, A., Grover, H., & Miller, C. (2008). Vulnerability and capacity: Explaining local commitment to climate change policy. *Environment and Planning C: Government and Policy, 26*, 544–562.

Part II

Climate Change Impacts on the Urban-Regional Level of Santiago de Chile

Downscaling Climate Changes for Santiago: What Effects can be Expected?

2

James McPhee, Gonzalo Cortés, Maisa Rojas, Lilian Garcia, Aniella Descalzi, and Luis Vargas

Abstract

This chapter describes the methodology used to analyse climate scenarios and their impact on hydro-meteorological variables in the Metropolitan Region of Santiago de Chile (MRS) and the results thereof. Using a downscaling methodology for future IPCC A2 and B1 scenarios (and B2 for stream flow), temperature, precipitation and secondary variable trends are estimated for the 2045–2065 time frame. The findings suggest that Santiago will be a drier and hotter city in the near future and have a high number of days with extreme temperatures. Lower precipitation rates are expected to lead to decreasing magnitudes in the stream flow of the two main rivers, Maipo and Mapocho, particularly in the summer months. Based on the data presented below, expected climate change impacts are analysed and adaptation needs identified for the MRS.

J. McPhee (✉)
Department of Civil Engineering, Faculty of Physical and Mathematical Sciences, Universidad de Chile, Santiago, Chile

Advanced Mining Technology Center, Faculty of Physical and Mathematical Sciences, Universidad de Chile, Santiago, Chile
e-mail: jmcphee@ing.uchile.cl

G. Cortés
Department of Civil Engineering, Faculty of Physical and Mathematical Sciences, Universidad de Chile, Santiago, CP, Chile

Department of Civil and Environmental Engineering, University of California Los Angeles, Los Angeles, CA, 90095, USA

M. Rojas
Department of Geophysics, Faculty of Physical and Mathematical Sciences, Universidad de Chile, Santiago, Chile

L. Garcia • A. Descalzi • L. Vargas
Department of Electrical Engineering, Faculty of Physical and Mathematical Sciences, Universidad de Chile, Santiago, Chile

K. Krellenberg and B. Hansjürgens (eds.), *Climate Adaptation Santiago*,
DOI 10.1007/978-3-642-39103-3_2,

Keywords
Climate change • Downscaling • Statistical analysis

2.1 Introduction

The current climate in the Metropolitan Region of Santiago de Chile (MRS) is characterized by warm dry summers and cold winters that concentrate most of the annual rainfall during the latter period. Average maximum temperatures in the summer range from 28 to 30 °C, while minimum temperatures during winter vary from 0 to 5 °C. Precipitation occurs primarily during the winter months of June, July and August, with annual precipitation ranging from 200 to 500 mm. Snow events, which are quite rare in the low elevation areas where the city of Santiago is located, dominate in high elevation Andean watersheds and serve as temporary water reservoirs for human, agricultural and industrial use. Trends in the last 30 years have indicated an increase in temperature for the upper elevation stations in particular, and a decline in precipitation amounts (Cortés et al. 2012). Although climate projections derived from models for the Latin American region show some variability in values, these trends correspond among the different simulations: warming temperatures and a reduction in winter precipitation throughout the continent, and less precipitation during all seasons along the southern Andes (Vera et al. 2006).

So far, climate change projections for the MRS have either been a component of large-scale studies (e.g., CEPAL 2009) or tailored made to sectorial needs (agriculture, water resources). This type of comprehensive information, however, is ill suited to analysing climate change impacts at urban-regional level (cf. Chaps. 4–7) or identifying adaptation needs. Hence the aim of this chapter is to present downscaled information for the MRS at a level of detail appropriate to outlining adaptation measures relevant to the urban environment. The chief benefits include the availability of daily scale projections to study the future variation of meteorological events, such as intense precipitation and extreme temperatures. At the same time, the absence of a dense network of spatially distributed meteorological data prevents a more detailed spatial analysis of the projected climate for the region.

This chapter builds upon previous studies performed at a national level in order to provide an in-depth analysis. The findings refer to the 2045–2065 period, enhancing previous works dealing with climate change projections for the end of the twenty-first century. These should be understood as an average for the entire period and not a prediction for each individual year. Where possible, they include a standard error bar. The latter represents the standard deviation that quantifies to a certain extent the uncertainty involved in future scenario models. Chile's unusual position between oceanic and continental climates demands that uncertainty estimates become standard in future studies on climate change impacts and adaptation.

The chapter is structured as follows: Sect. 2.2 provides information on the data used to analyse climate change projections for the MRS. Section 2.3 gives a detailed description of the methodologies applied to downscale climate projections to scales appropriate for adaptation studies. Section 2.4 presents the principal results for individual variables in the future scenario analysis (2045–2065) in the context of the historical records on which the climate change impact analysis is based. Section 2.5 derives key conclusions and discusses the methodological constraints and uncertainty issues involved in using these figures for further analysis on climate change impact and adaptation needs.

2.2 Data Sources and Methodological Steps

In order to project future climate changes for the MRS, Global Circulation Models (GCMs) simulating the Earth's climatic system are adapted or downscaled to the appropriate spatial scale (for further detail, see Sect. 2.3). This allows for the assessment of future climate changes for climate-relevant variables such as temperature, number of hot days or precipitation for the 2045–2065 time frame, the earliest period for which GCM data is available after the baseline time frame (1960–1999). Future GCM-based climate projections are based on scenarios describing the evolution of greenhouse gas emissions. Two main scenarios, i.e., SRES A2 and B1, are assessed in this contribution, while the B2 scenario is explored for stream flow projections (for details, see IPCC 2000, 2001, 2007). The next step is to average cross-model results and compute standard deviations for overall time-frame averages. This provides uncertainty estimates due to GCM variability. In addition to the statistical downscaling of global models, analysis of the principal climate variables allows the assessment of historical and current climate conditions, as well as expected future climate changes in the Metropolitan Region of Santiago de Chile. These include temperature and precipitation as primary data. Secondary data is comprised of glacier development, the shift in the isotherm 0 °C, water run-off, wind velocity and insolation. The historical analysis is based on data obtained from meteorological stations owned and operated by public agencies. Many of the stations in the MRS used for the purpose of this study have records that date back 40 years (1970–2010). There are, however, major information gaps in the records of some stations.

Figure 2.1 shows the stations used for precipitation and temperature analysis. The light grey area indicates the limits of the MRS, while the darker area is an approximation of current urban limits. Stations were selected by analysing the temporal range of the data and the consistency of measured values. Table 2.1 provides further information on the stations.

2.2.1 Precipitation

The time span used to analyse historical precipitation averages corresponds to the concurrent period between the GCM models and the observed data (1970–2000 where available) (Table 2.1). Uncertainties remain with regard to spatial representativeness of point measurements and measuring methodologies adopted by the institutions in charge of station maintenance. Since no evident errors were observed, no data correction was performed.

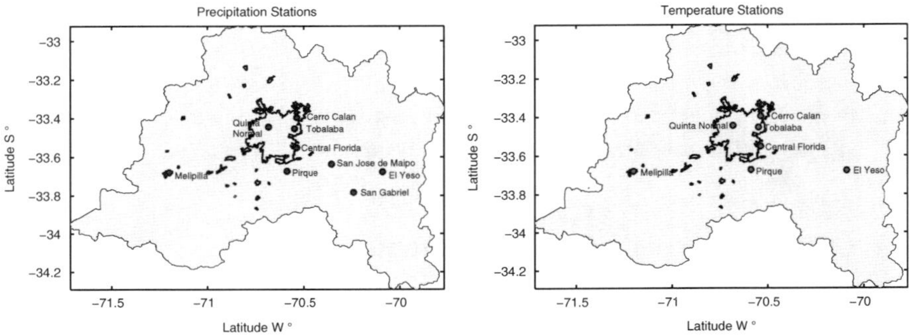

Fig. 2.1 Climatological stations (**a**) Precipitation data, (**b**) Temperature data (*Source*: Authors' own presentation)

Table 2.1 Meteorological stations (*Source*: Authors' own presentation)

ID	UTM N	UTM E	Elev. (m)	Beginning of record	End of record	Variables recorded
Central Florida	6286445	357350	770	1977	2010	Daily precip., mean daily temp., max/min daily temp.
Melipilla	6271064	296074	168	1972	2010	Daily precip., mean daily temp., max/min daily temp.
Pirque	6272845	352877	659	1968	2010	Daily precip., mean daily temp., max/min daily temp.
Quinta Normal	6298113	343588	527	1977	2007	Daily precip., mean daily temp., max/min daily temp., radiation, relative humidity, wind
San José de Maipo	6277311	374507	964	1972	2010	Daily precip.
Tobalaba	6297259	356163	652	1979	2010	Daily precip., mean daily temp., max/min daily temp.
Cerro Calan	6303810	357081	848	1976	2010	Daily precip., mean daily temp., max/min daily temp.
El Yeso	6273104	399083	2475	1963	2010	Daily precip., mean daily temp., max/min daily temp.
San Gabriel	6261211	385240	1266	1978	2010	Daily precip.
Pudahuel	6304000	333900	480	1980	2005	Radiation, relative humidity, wind

Precipitation values generally tend to increase according to altitude (Cortés et al. 2011). While precipitation may fall as snow above 2,000 m.a.s.l. during winter, in the lower lying areas of the city of Santiago (below 1,000 m.a.s.l.) only isolated events of solid precipitation occur every few years. These snow events are not recorded or differentiated by the stations due to instrumental constraints, and snow usually remains on the ground for a few hours only. Precipitation gradients observed are due to the orographic effect of the Andes Cordillera on frontal systems coming from the Pacific Ocean, the result of which is a major increase in precipitation when stations away from the Cordillera are compared with those located in the foothills or on elevated sites (Cortés et al. 2011).

All of the stations analysed present a similar regime in terms of precipitation frequency distribution, with a high number of "dry" days or days with less than 1 mm precipitation, and—depending on its intensity—a decline in the number of days with higher precipitation. Due to its location at a higher altitude, the El Yeso station shows evidence of a greater number of days with precipitation than other stations. The remaining stations present similar regimes, each with an annual total of approximately 30 days with precipitation. For the present analysis, the 1980–2000 period was used, as it was the common period with the smallest number of gaps.

Trends in precipitation have been examined in various studies (CONAMA 2006; Quintana and Aceituno 2006) and their common observation is a slight decreasing trend for central Chile, with lower total precipitation amounts. Considering the precipitation gradients, the stations for this study show varying precipitation values, ranging on average from 345 mm p.a. at Quinta Normal to 657 mm p.a. at San Gabriel. Precipitation patterns are similar for all stations, with higher precipitations from May until August and lower precipitation in the summer (December to February). Standard deviation is high, however, indicating that precipitation is characterized by high variability. One reason for irregular precipitation could be the influence of ENSO phenomena in the MRS, which introduces high variability during warm events in the shape of higher precipitation amounts ("El Niño") and low precipitation amounts during cold events ("La Niña") (Cortés et al. 2011).

2.2.2 Temperature

The historical temperature analysis is based on data from seven meteorological stations presented in Table 2.1. The number of stations available is small: a dense climatological network has not been established, and only one station, Quinta Normal, is located within the urban limits of Santiago. Cerro Calan, Tobalaba and Central Florida are suburbs in low-density areas. Spatial interpolation between the stations, which could serve to generate area-wide information for further analysis, was avoided, since the interpolation of point measurement values has severe limitations and adds no significant information to this particular study. The El Yeso station located in the higher Andean area, for example, would affect temperature interpolation in the entire northeast region as a result of its altitude (Table 2.1).

The time span of temperature records varies from station to station but covers most years between 1970 and 2000. The seven station locations are heterogeneous. Their individual features determine the representativeness of their detailed meteorological conditions. Quinta Normal station is located in a park in the centre of Santiago, for example, while Cerro Calan is situated on a hill in the eastern suburbs. Although geographical location should play a role in determining the spatial distribution of temperature, there is not enough evidence to show that any single predictor (such as elevation, as in the case of precipitation) can explain its spatial variability. The only station with obvious temperature differences as a result of elevation is El Yeso, which is located in the Andean Cordillera (Cortés et al. 2012).

Maximum annual temperatures for the stations in the Maipo Basin vary only slightly, and range on average from 21.7 to 22.9 °C. The sole exception here is the El Yeso station, diverging significantly with an annual average maximum temperature of 13.7 °C. For all stations, maximum temperatures during summer months normally exceed 25 °C. Some stations even report average maximum temperatures of more than 30 °C for January and February (e.g., Cerro Calan and Central Florida). Maximum winter temperatures reach an average of 15 °C, with the exception of El Yeso with a mere 5.9 °C maximum average temperature in July. The standard deviation is quite low for the stations in the Maipo Basin, with a slight increase for the more elevated El Yeso station.

Minimum temperatures during summer months range on average from 11 to 14 °C for stations in the Maipo Basin, compared to 8.7 °C for El Yeso. Minimum temperatures during the winter drop to an average of 4 to 6 °C in Melipilla, Quinta Normal, Tobalaba, Central Florida and Cerro Calan, with an average of merely 1.7 °C in Pirque. El Yeso has 4 months of average minimum temperatures below zero.

Tables 2.2 and 2.3 indicate the number of days for each year that show maximum temperatures above 30 °C and minimum temperatures below 0 °C, along with standard deviation computed from the time series of each variable.

Falvey and Garreaud (2009) analysed historical temperature data in central Chile (27.5–37.5°S) for the 1960–2006 period and found positive trends from 1975 to 2006 in stations located in the Andes and in the central valley, the location of the MRS. Cooling patterns were observed for coastal and low-lying stations. Daily maximum and minimum temperatures increased proportionally and in accordance with the elevation of the measuring station. The key conclusion of their study is the presence of a warming trend in the Andean regions of central Chile and—somewhat less significant but still observable—a similar trend in the valley regions.

2.2.3 Stream Flow

Stream flow data for the two main rivers in the MRS is obtained from two stations: Los Almendros measures the Mapocho river runoff, while the San Alfonso station measures the Maipo River stream flow. The location and elevation of these stations is presented in Table 2.4. Average, minimum and maximum monthly run-off values are given for each month. The analysis of the historical period included calculating exceedance probabilities (PEXC) based on the 1960–2000 period, which is more accurate than merely calculating minimum or maximum values, as it allows for a proper analysis of stream flow distribution throughout the year. If a value of X m^3/s corresponds to an exceedance probability of 60 % for a certain month, for example, then 60 % of the monthly average stream flow measured during that 40-year period

Table 2.2 Number of days with maximum temperatures above 30 °C (*Source*: Statistical data from meteorological stations)

	Cerro Calan	Quinta Normal	Pirque	Melipilla	Tobalaba	Florida
Average no. of days per year	67.92	54.58	30.48	17.00	44.56	55.50
Standard deviation	10.58	8.59	12.78	7.83	11.09	17.25

Table 2.3 Number of days with minimum temperatures below 0 °C (*Source*: Statistical data from meteorological stations)

	Cerro Calan	Quinta Normal	Pirque	Melipilla	Tobalaba	Florida
Average N° of days per year	0.81	7.04	28.54	2.31	5.92	4.54
Standard deviation	1.23	4.23	14.53	2.59	4.32	3.22

Table 2.4 Stations recording stream flow data (*Source*: Dirección General de Aguas)

ID	UTM N	UTM E	Elev. (m.a.s.l.)	Ti	Tf*	DT
Mapocho in los Almendros	6307006	365534	1,024	1961	–	Day and month
Maipo in San Alfonso	379641	6266823	1,108	1961	–	Day and month

*Final or ending time of the time series

is above X m^3/s. All of the data presented consists of monthly values, so that extreme run-off events may not be represented.

The analysis of stream flow projections shows considerable variability in run-off values for the historical period. Stream flow in the Maipo River measured as high as 413.5 m^3/s in January at the San Alfonso station, for instance, but also as low as 35.5 m^3/s. However, these are extremes. Using exceedance probabilities, the results indicate that it can be expected that 90 % of the time monthly average stream flow exceeds 74.0 m^3/s in January, while run-off values over 272.6 m^3/s occur with a probability of only 10 %. Stream flow values are generally lower in the winter. Although monthly flow exceeds 22.3 m^3/s 90 % of the time in June, for example, it exceeds 52.5 m^3/s in only 10 % of cases. The minimum for this month is 11.3 m^3/s, the maximum 118.7 m^3/s. Data for all 12 months of the year with the specific exceedance probabilities for the Maipo River are shown in Table 2.5.

The high variability of stream flow values for the Maipo River is even more striking when transferred to a graphic chart (Fig. 2.2). The San Alfonso station for the Maipo River shows a prevalence of spring and summer high flows. Known as "snow dominated", this type of regime indicates that most of the stream flow comes from the melt of snow and ice accumulated during the winter months. It is important for a Mediterranean climate type, as most of the water flows during the dry months, allowing for irrigation and water supplies during the hot and dry season. In a climate change scenario, changes in the amount and timing of stream flow should be expected depending on the type of change predicted. Hotter temperatures in winter and spring, for example, would result in a shift in the peak flow and thus an

Table 2.5 Stream flow values of the Maipo River in San Alfonso in m^3/s based on monthly average data for the 1961–2000 period (*Source*: Dirección General de Aguas)

PEXC	APR	MAY	JUN	JUL	AUG	SEP	OCT	NOV	DEC	JAN	FEB	MAR
90 %	31.1	21.8	22.3	21.6	19.6	28.6	38.4	61.8	74.0	69.4	54.3	45.3
80 %	36.7	28.9	26.7	26.0	25.4	31.8	46.8	76.7	93.6	75.5	64.4	49.8
70 %	38.0	31.1	28.8	28.6	27.4	33.6	54.7	87.7	110.4	85.2	66.4	50.8
60 %	39.6	34.4	31.9	30.1	30.5	36.3	58.5	98.8	126.5	88.8	71.3	57.0
50 %	42.8	35.6	34.4	33.4	32.2	37.9	61.0	108.6	140.8	105.1	81.6	59.7
40 %	47.5	40.3	40.5	35.4	36.6	43.4	67.6	127.7	152.0	145.8	94.9	69.0
30 %	49.3	44.4	42.3	40.9	42.4	48.4	76.5	133.6	178.6	167.6	115.2	72.4
20 %	59.9	50.1	48.9	48.8	46.9	56.4	87.3	139.8	222.4	207.3	134.3	78.8
10 %	74.2	60.3	52.5	55.3	55.8	71.1	93.7	158.3	281.5	272.6	154.9	106.6
Minimum	19.6	12.0	11.3	18.7	15.4	9.4	31.1	39.5	28.6	35.5	43.5	24.2
Maximum	94.3	199.4	118.7	71.2	65.0	89.7	123.3	206.4	373.3	413.5	278.8	163.1

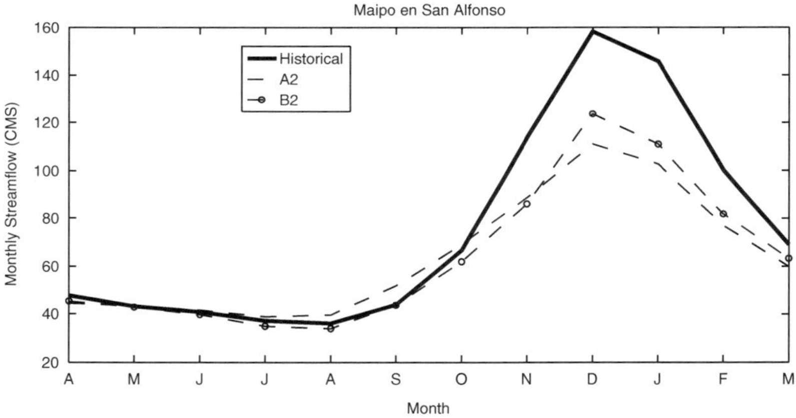

Fig. 2.2 Mean monthly stream flow projections for the Maipo River at San Alfonso watershed (*Source*: Authors' own presentation)

earlier occurrence of peak stream flow, while lower precipitation amounts would lead to lower volumes of water during the entire season.

The second river under review is the Mapocho River. Results for this station are presented in Table 2.6.

Unlike the Maipo River, the Mapocho River shows significant flows in spring and late winter (Table 2.6), as well as during wet periods in early winter. It could eventually be more sensitive to warming temperatures, as its watershed is lower than that of the Maipo River. Increasing temperatures will expose a larger area of the watershed to liquid precipitation rather than snow, and subsequently have a heavy impact on stream flow. It implies, on one hand, earlier peak stream flow values and, on the other hand, a decrease in the thickness and extent of the snow-covered area in the basin. The impact of a rising isotherm 0 line is also significant,

Table 2.6 Stream flow values for the Mapocho River in los Almendros in m^3/s based on monthly average data (*Source*: Dirección General de Aguas)

PEXC	APR	MAY	JUN	JUL	AUG	SEP	OCT	NOV	DEC	JAN	FEB	MAR
90 %	1.0	1.0	0.5	0.7	0.8	0.9	2.0	1.9	2.8	2.7	2.3	1.5
80 %	1.3	1.4	0.8	1.1	1.2	2.2	3.3	3.3	3.1	3.3	2.5	1.9
70 %	1.5	1.5	1.6	1.9	2.1	3.0	4.8	5.7	4.7	4.3	3.1	2.1
60 %	1.7	1.7	1.8	2.3	3.4	4.8	6.5	7.1	6.4	5.0	3.6	2.2
50 %	2.0	2.0	2.3	3.2	3.7	5.4	7.5	8.3	8.9	6.2	4.0	2.6
40 %	2.3	2.2	2.9	3.8	5.0	7.5	12.2	13.1	11.2	7.1	4.8	3.2
30 %	2.7	2.6	3.6	4.7	6.0	9.0	15.4	15.5	13.6	8.0	5.6	3.7
20 %	2.8	3.2	5.2	5.9	7.1	11.7	18.7	19.0	16.5	9.9	6.2	4.5
10 %	3.6	3.7	8.3	9.3	8.6	14.0	21.1	22.8	28.9	17.5	7.6	5.3
Minimum	0.8	0.7	0.4	0.2	0.3	0.3	1.0	1.0	1.5	2.3	1.5	1.1
Maximum	8.0	18.5	22.7	22.3	19.3	23.3	25.1	41.8	38.7	26.0	15.2	6.1

since the presence of lower elevations in the basin implies that a greater area contributes to runoff in the course of each storm event.

2.2.4 Glaciers

Although glaciers in the region (e.g., the Echaurren Glacier) have been studied for the past decades (e.g., Casassa 1995), current data for the MRS is confined to an updated glacier inventory that characterizes the glacier-covered area, but not the glacier volume or its dynamics (Rivera et al. 2000, 2002). Neither does a recently developed monitoring network for the MRS reveal trends for the evolution of these glaciers. Hence this chapter will analyse trends and the state of the art for regional ice masses based on available publications and local research efforts. Clearly, the topic of glaciers and their evolution under climate change in the central Andes poses several—as yet unanswered—interesting questions, and more research is needed to improve quantification of their current contribution to water resources and their possible future evolution.

Glaciers and permanent snowfields in the Andes Cordillera store large quantities of fresh water. It is estimated that they contain a total water equivalent volume of about 30.6 km^3 (Garín 1986). Glaciers are therefore crucial to the hydrological cycle and the water supply for the Metropolitan Region of Santiago. According to a glacier cadaster performed by the Dirección General de Aguas (DGA), there are 647 glaciers in the central area of Chile (Marangunic 1979). Their surface adds up to almost 422 km^2, with the mean glacier area equal to 0.65 km^2 (Garín 1986). Recent studies performed for the region show consistent trends in glacier retreat for the Andean mountains. Le Quesne et al. (2009) calculate that the complete glacier surface in the central Andean region dropped by approximately 3 % between 1955 and 2000. Detailed studies report for some glaciers an even more pronounced loss.

The Aconcagua River basin glaciers, for example, have experienced a 20 % area reduction on average since 1955 (Bown et al. 2008). The Juncal Norte glacier located in the headwaters of the Aconcagua River, approximately 70 km northeast of Santiago, shows a retreat of about 50 m per year (Rivera et al. 2002). On the other hand, the Dirección General de Aguas (DGA) has recently published a report on the historical trends of mass balance in the Echaurren Norte glacier. The Echaurren Norte drains to the Laguna Negra in the headwaters of the Volcán River, itself a tributary of the Maipo River. With a surface area of merely 0.226 km^2 (DGA 2010), it is one of the most thoroughly studied glaciers in the region. Several field campaigns show that the net accumulated mass balance of the glacier was relatively stable until approximately 1991, when it experienced a sudden and steep drop that lasted until 2000. Having recovered to a certain extent, it is now stable. The entire period 1975–2008 shows evidence of a total accumulated loss in glacier mass balance of almost 8 m water equivalent[1] (DGA 2010).

2.2.5 Other Secondary Variables

Records for secondary variables such as wind velocity or insolation were obtained from the two meteorological stations that measure them: Pudahuel and Quinta Normal. Table 2.7 presents information on these two stations. The seasonal averages and standard deviation values for each of the variables are presented in Table 2.8. With only two stations available for analysis, no spatial interpolation was performed, as results would prove unreliable.

2.3 Downscaling of Future Climate Scenarios: Methodological Explanations

Global Circulation Models (GCMs) are mathematical models that simulate the Earth's climatic system. Land, ocean and atmospheric processes are represented by calculations performed on grid divisions of the Earth's surface, each with resolutions of up to hundreds of kilometres. For climate change impact studies, GCM simulations must therefore be downscaled to the spatial scale relevant to the study concerned. In this specific case, the relevant spatial scale is the MRS, which spans the entire width of the country (approximately 200 km) and may be of similar scope to most GCM grid elements. The uncertainty related to model performance and structure can be overcome by taking several GCMs into account. In this study we use 10–15 different models for any one variable or indicator of climate change (e.g., the average annual precipitation for the 2045–2065 time frame). Cross-model averages are then calculated, indicating standard deviations computed for overall

[1] In water equivalent calculation, the density is assumed to be 0.9 g/cm^3. One meter of ice is therefore approx. 90 cm water equivalent.

Table 2.7 Secondary variables at two meteorological stations (*Source*: Authors' own presentation)

ID	UTM N	UTM E	Elev. (m.a.s.l.)	Ti	Tf	DT
Pudahuel	6304000	333900	480	1980	2005	Seasonal
Quinta Normal	6298113	343588	527	1980	2005	Seasonal

Table 2.8 Secondary variable climatology at selected stations (1980–2005 period) (*Source*: Dirección Meteorológica de Chile)

			Season (months)				
Variable	Units	Station	DJF	MAM	JJA	SON	Annual
Wind velocity	(km/h)	Pudahuel	13.4 (1.5)	7.9 (1.2)	5.6 (2.1)	10.2 (1.8)	9.7 (1.5)
		Quinta normal	6.9 (1.1)	3.7 (0.9)	1.6 (0.8)	5.7 (1.0)	5.3 (1.2)
Mean radiation	(Wh/m^2)	Pudahuel	315.5 (9.3)	163.2 (21.1)	96.6 6.8)	237.7 (11.9)	201.9 (11.1)
Relative humidity	(%)	Pudahuel	48.7 (11.9)	63.2 15.4)	74.9 (18.0)	61.1 (15.1)	62.0 (15.0)
		Quinta normal	51.7 (9.4)	66.3 (11.6)	76.6 (13.4)	62.6 (11.3)	64.3 (11.2)

Standard deviation from the mean is shown in parenthesis

time-frame averages rather than for year-to-year annual values. This allows for distinction between model uncertainty and inter-annual variability. The following models were used for the downscaling procedure: the CGCM3 model from the Canadian Center for Climate Modeling (CCCMA, Canada), the CM3 model from the Centre National de Recherches Météorologiques (CNRM, France), the AOM, E_H and E_R model from the Goddard Institute for Space Studies (GISS, USA), the CM4 model from the Institute Pierre Simon Laplace (IPSL, France), the MIROC_MEDRES model from the National Institute for Environmental Studies (NIES, Japan), the MK model from Australia's Commonwealth Scientific and Industrial Research Organization (CSIRO, Australia) , the ECHAM model from the Max Planck Institute for Meteorology (MPI, Germany) and the ECHAM model from the Istituto Nazionale di Geofisica e Vulcanologia (INGV, Italy).

2.3.1 Downscaling Temperature Projections

Downscaling techniques are required to establish a relationship between the data acquired from GCMs and the values observed at meteorological stations, since large-scale climatic patterns and regional features both influence regional climatic conditions (von Storch 1999; Wilby et al. 2004). Statistical downscaling is the "development of quantitative relationships between large-scale atmospheric variables and local surface variables" (Wilby et al. 2004). First, we create a statistical model that relates large-scale variations to local meteorological observations. The GCM model outputs are subsequently applied to a number of scenarios with the relationship found. The major advantage of this technique is its relative simplicity, making it "computationally feasible" for most studies. One obvious disadvantage, however, is the assumption that present day (or historical) relationships are valid for future scenarios. This can lead to invariances when the models reproduce seasonal variability for the region correctly (e.g., cold winters and hot summers), but day-to-day averages and short-term variability do not coincide with the observed data. This "time invariance" could have been ignored if several historical periods had been used to validate the relationship found; since the long-term observational data set required for validation would exceed the scope of this study, it was not applied.

2.3.2 Downscaling Precipitation Projections

Precipitation values in GCM models could be biased when compared to those observed for the MRS. In terms of frequency distribution, several GCM precipitation estimates show strong differences with respect to the observed climate. An attempt to correct or downscale these estimates would not produce valuable information on future climate since the differences come from limitations such as sub-adequate representation of the Andes Cordillera or ocean land coupled within the model pixel that represents the study region. Hence, the precipitation analysis made use of a subgroup of the available GCMs: CNRM_CM3, CSIRO_MK, MIROC_MEDRES and MPI_ECHAM.

While temperature data calls for measurement of only one variable, i.e., temperature, two variables must be downscaled for precipitation data prior to analysis: the amount or intensity of precipitation and wet-day frequency. The downscaling procedure must consider both variables to achieve a realistic representation of the precipitation distribution registered in each measuring station. Schmidli et al. (2006) propose a method for daily precipitation downscaling that was deemed adequate for this analysis. The methodology considers GCM modelled precipitation data as a predictor for observed precipitation data, as proposed by Widmann et al. (2003). The method consists in downscaling wet-day frequency and the precipitation intensity for each day: wet-day frequency is downscaled and the precipitation intensity subsequently adjusted so that the observed and the modeled series have the same daily precipitation intensity distribution. Application of this methodology does not require a specific climate. The idea is to correct potential biases in the distribution of climatic variables between models and observations and thus to preserve climatic patterns as distinct from absolute values. Since the methodology entails sequential downscaling—the separate adjustment of wet-day frequencies and precipitation magnitudes—the singularities of daily precipitation distribution in Mediterranean climates (highly asymmetric density functions, large proportion of zero-rainfall) are taken into account. Schmidli et al. (2006) obtained good results with this methodology in comparison to other downscaling methods.

2.3.3 Stream Flow and Secondary Data Projections

Stream flow data for future scenarios was obtained with a hydrological model, WEAP21. Based on hydrological catchments, this is a semidistributed model where each catchment may contain different physical and climatic properties. For the purpose of exploring future values of stream flow, the model was first calibrated against historic data from meteorological stations: monthly average stream flow data at Maipo in San Alfonso station (the model is calibrated using a monthly scale), monthly precipitation at San Gabriel station and the monthly mean temperature at Pirque station. Future scenarios derived from GCM data were downscaled to these stations. Projected future downscaled precipitation and temperature values were used as input for the hydrological model to obtain output for future scenarios. Data for future scenarios on stream flow was available for the IPCC A2 and B2 scenarios only. Although the B2 scenario is not identical to the B1 scenario, both scenarios present an "optimistic" point of view on future emissions and up to the year 2050 are practically indistinguishable. Although detailed data sets of the results are available, for clarity purposes this chapter will only present the average values for future scenarios.

For variables such as wind velocity, radiation and relative humidity, downscaling is much more uncertain, as these variables are correlated to a regional climatic pattern as well as to local singularities. Radiation measurements will be affected, for example, if the location of the station is prone to fog formation, and wind velocity measurements modified if there exist hills or buildings in the immediate vicinity of the station. As GCM resolutions cannot resolve these effects, model projections of secondary variables will probably be far from representative of measurements performed at a point level. The statistical approach adopted for temperature was nevertheless used for the secondary data, taking monthly measurements of these variables into account. However, since the uncertainty derived from the importance of local effects on these variables is high, results for secondary variables (especially wind, relative humidity and radiation) must be seen differentially; in other words, the difference between the scenarios (historic, A2 and B1) must be taken into account, not the absolute value of the variable projected by the GCMs.

2.4 Future Climate Analysis Results

This section presents the key findings on future predictions for the 2045–2065 time frame and discusses them for each climate variable.

2.4.1 Temperature

Figures 2.3 and 2.4 show future projections of maximum and minimum daily temperatures for each month of the year and for each meteorological station. Where available, observed (i.e., historical) data refers to the 1970–2000 time

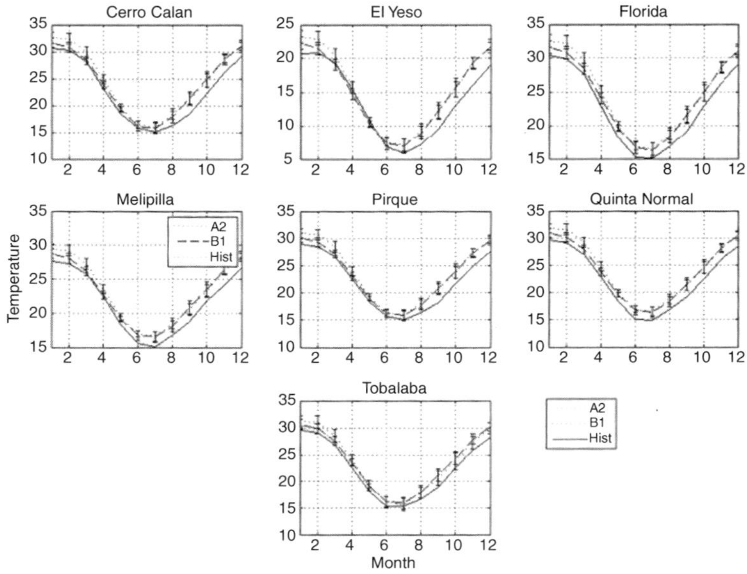

Fig. 2.3 Monthly maximum daily temperature (historical and projected (2045–2065) values) (*Source*: Authors' own presentation)
Note: Error bars show the annual standard deviation in projected future values due to GCM differences

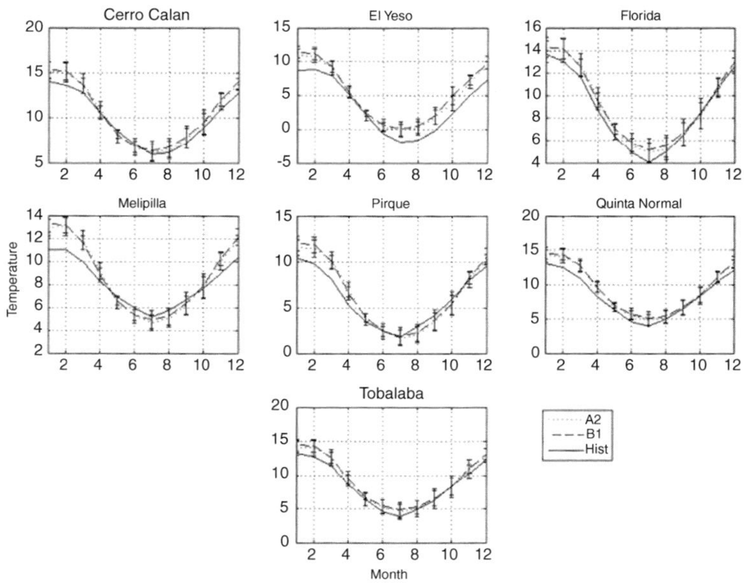

Fig. 2.4 Monthly minimum daily temperature (historical and projected (2045–2065) values) (*Source*: Authors' own presentation)
Note: Error bars show the annual standard deviation in projected future values due to GCM differences

span. Future projection averages are based on the A2 and B1 greenhouse gas emission scenarios for the 2045–2065 time period.

Each of the seven stations experiences a rise in annual mean temperature in the order of 1.5 °C in the two scenarios analysed for the 2045–2065 period. The increase in temperature is consistent for every month but appears to be more significant for summer than winter months, i.e., the MRS will be particularly affected between November and March, as higher temperatures will produce more heat waves and greater extremes in temperature (Fig. 2.3). This observation should be backed up with the total number of days with maximum temperatures above 30 °C (results presented in Fig. 2.5).

The analysis of average minimum temperatures shows that differences between the stations are far greater here than in the case of maximum temperatures. While Central Florida reports an annual minimum temperature of 12.6 °C on average, the other stations of the Maipo Basin reach an average of no more than 8 °C: Pirque with an average 6.0 °C—the lowest value for this group of stations—and Cerro Calan averaging 9.9 °C in between. Again, due to its elevation, El Yeso shows the lowest minimum temperature of all stations with an average 3.6 °C (Fig. 2.4).

Minimum temperatures also show a consistent increase, albeit less than that encountered for maximum temperatures. Since the former are high compared to those in rural stations such as Pirque, climate change impact will not be clearly visualized in minimum temperatures within the limits of the city. The change in minimum temperatures is still positive: more than one degree for most of the stations under review for the two scenarios. This shift is less significant in the spring months, and uncertainty is also on the increase during this period. For summer and winter, however, the warming signal is strong (Fig. 2.5).

Under climate change projections, all stations would experience a total annual increase in the number of days with maximum temperatures above 30.0 °C. Some of these changes are significant. The Santiago station (Quinta Normal), for example, shows an increase of more than 30 days with extremely high temperatures; this also applies to the Cerro Calan, Florida and Pirque stations for the more severe A2 scenario. Even the optimistic B1 scenario indicates a significant increase in the number of days with extreme temperatures for all stations. This could seriously affect the quality of life in Santiago in the summer time. The result is consistent across all GCMs included in the analysis.

Without exception the stations present a smaller number of days with freezing temperatures. This is especially significant for those that presented a higher number of days with freezing temperatures in the past, such as the Pirque and El Yeso stations.

2.4.2 0 °C Isotherm Altitude

According to a study performed by Carrasco et al. (2005), a change of approximately 150 m (positive increase) was observed for the isotherm-0 line between 1975 and 2001 in central Chile, and subsequently a shift in the snow line and

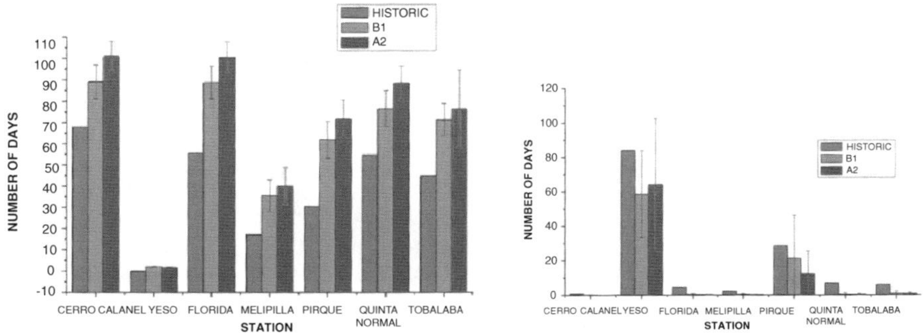

Fig. 2.5 Annual number of days with temperatures above 30°C (*left*) and below 0°C (*right*) for each station (for the historical scenario and the two future scenarios (A2 and B1 for 2045–2065)) (*Source*: Authors' own presentation)
Note: Variations in the number of days with maximum temperatures above 30°C are due to the different downscaling relationships derived for each station; some stations have different historical values and local effects on temperature

equilibrium line. Data for this study was obtained with radiosonde measurements (not available to the public). For a station located near the Pacific Ocean at the same latitude as Santiago, the mean annual isotherm line was located at 3,500 m.a.s.l. approximately, varying between values of 4,250 m.a.s.l. in the summer and 3,000 m.a.s.l. in the winter. Given that a station located near the ocean should exhibit warmer temperatures than those observed inland during winter, the values obtained, based on the Pirque temperature record and the mean gradient, are deemed to be close to those observed. Hence future temperature scenarios will consider Pirque and the same gradient as a baseline for understanding isotherm 0 line changes caused by warming. The analysis of the isotherm 0 position is vital as it indicates when heavy precipitation events will have a strong impact on the city. Due to the high position of the isotherm 0 line, storm events in March or April can lead to increased flooding, because of greater amounts of liquid precipitation available.

Due to uncertainty in the temperature gradients, however, the low number of stations and the fact that each storm has its own isotherm 0 line position, changes addressed in this chapter must be taken into account in qualitative rather than absolute terms. The position of isotherm 0 lines is important as it determines the division between liquid and solid precipitation in the upper elevations. A higher isotherm 0 line would bring increased liquid precipitation areas for each storm, with greater flooding and a higher amount of sediments carried during each event. Table 2.9 presents the calculated isotherm derived from temperature gradients for the future scenarios. These results were calculated indirectly from mean monthly temperature projections for Pirque station. These monthly values were calculated at the same time from mean maximum and minimum temperature projections for each month.

Table 2.9 0 °C isotherm variations due to climate change (*Source*: Authors' own presentation)

0 °C Isotherm Altitude (m.a.s.l)			
Month	Historic	A2	B1
January	4,235	4,576	4,525
February	4,197	4,627	4,520
March	4,001	4,442	4,278
April	3,322	3,695	3,527
May	2,673	2,841	2,777
June	2,295	2,413	2,384
July	2,153	2,292	2,278
August	2,241	2,384	2,340
September	2,475	2,671	2,670
October	2,929	3,106	3,078
November	3,514	3,787	3,741
December	3,978	4,276	4,244
Annual	3,146	3,398	3,334

As they are approximate and based on one station only, these values must be seen in perspective (i.e., a comparison of the differences in historic and future scenarios, and not their absolute values). El Yeso, for example, shows a mean temperature above freezing point for the A2 and BI scenarios, but isotherm results indicate that freezing temperatures could still prevail at the altitude of El Yeso

2.4.3 Precipitation

Future precipitation projections were obtained by downscaling data from the different GCMs available to each station in the historical scenario. These were compared to values downscaled for the A2 and B1 future scenarios for the time period 2045–2065. Precipitation model outputs differ for all scenarios (A2, B1, observed and historic). The models used were those closest to the Mediterranean characteristic of the climate in central Chile. Figure 2.6 presents the results for annual precipitation at each station and for each scenario.

Both scenarios show a distinct reduction in precipitation amounts for almost all months of the year. This is particularly significant for the A2 scenario, where reductions of between 10 and 30 % are predicted by the model ensemble mean (20–100 mm less precipitation each year, depending on the station in question). It is important to clarify that these differences are relative to the downscaled ensemble mean, i.e., the historical values downscaled from the climate models, not those observed at individual stations.

With regard to precipitation intensity, Fig. 2.7 presents the precipitation categories for each station, derived from the ensemble mean. The decline in the number of days with precipitation for each category coincides with the general decline in precipitation expected for the region. Yet, this decrease is not as clear for the higher intensities: most precipitation in the region is expected to come from

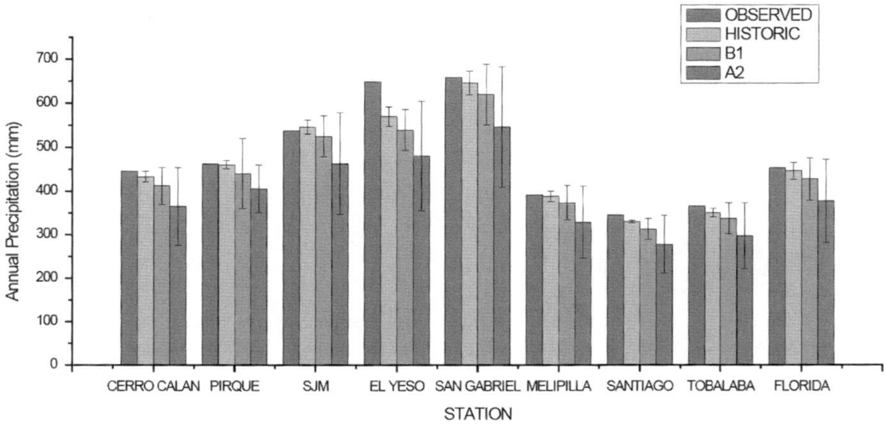

Fig. 2.6 Total annual precipitation changes (*Source*: Authors' own presentation)
Note: Historic precipitation is derived from models, whereas observed precipitation is obtained from meteorological stations (observed climatology). Two factors explain the differences between "historic" and "observed": the downscaling methodology is unable to reproduce observed precipitation faithfully and models may have difficulty in reproducing local effects on total annual precipitation. The bars represent standard deviations found in the models

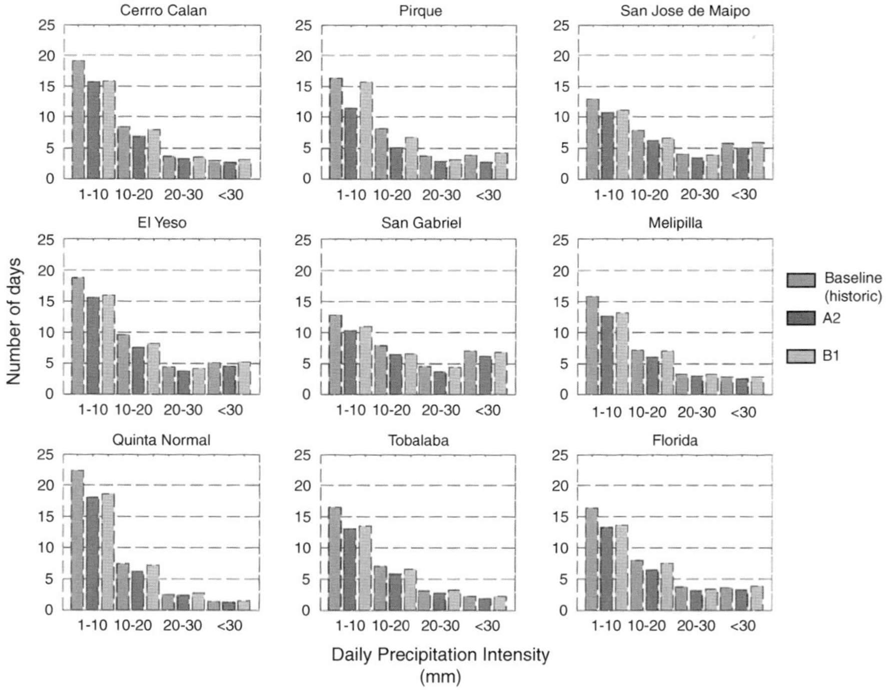

Fig. 2.7 Historic and future (2045–2065) precipitation for each analysed station grouped by scenario and daily intensity (*Source*: Authors' own presentation)
Note: Each group of three columns represents a specific daily precipitation intensity

high-intensity storms, since the overall decline in precipitation is primarily signified by a decrease in the number of days with low to medium precipitation amounts and a major decline in the number of days with 1–10 mm of precipitation. No significant changes are projected for the number of days with high-intensity precipitation, or for the average intensity of these events. Despite various sources of uncertainty, most of the models agree that Santiago will be a drier city in the future, with lower amounts of annual precipitation due to fewer rainy days.

2.4.4 Stream Flow

For the future period under analysis, dramatic changes can be observed in the hydrologic regime of the rivers concerned. Peak stream flow dates will shift to earlier months, and decreases of up to 40 % are observed for some summer months. Peak flow in winter will increase, as higher temperatures bring earlier melting of snow and ice. During the winter, lower precipitation rates and higher temperatures will reduce the amount of snow that accumulates in the high Andes, leading to a further reduction in run-off during spring and summer. A further crucial impact indicated in the hydrological model is the presence of higher stream flow values in autumn and winter. Considering model results show lower precipitation occurrence in the future, this increase in run-off is probably solely the result of increased melting during these months and a higher isotherm 0 line. Results are presented as averages for the 2045–2065 period (Table 2.10).

2.4.5 Glaciers and Secondary Data

As glaciers account for a significant percentage of stream flow volumes in February and March, the reduction in volume could have implications for water resource availability in the region during the summer months. The specific contribution of glacier melt to overall water availability in the Metropolitan Region is unknown as yet, however, with possible estimates ranging from 30 to 67 % for rivers such as the Maipo (Peña and Nazarala 1987). The exact amount of this contribution has yet to be quantified, and most glacier dynamics in the region must be examined exhaustively if meaningful estimates of future glacier evolution and its impact on water resource availability are to emerge. It is worth mentioning the special status glaciers in the central parts of Chile possess. According to Borquez et al. (2006), glaciers in the central Andes Cordillera can be classified as "polithermal". They are found at relatively low altitudes (5,000 m.a.s.l. and below), with the ablation zone occasionally located at very low elevation. During the summer, the lower sections of these glaciers experience positive temperatures, producing run-off especially during the hot months. In a climate change context, these glaciers could be affected more than others, since major glacial mass areas here are vulnerable to high temperatures.

Table 2.10 Average monthly values for the Maipo River at San Alfonso and Mapocho River at Los Almendros stream gages (*Source*: Authors' own presentation)

SCENARIO	APR	MAY	JUN	JUL	AUG	SEP	OCT	NOV	DEC	JAN	FEB	MAR
Maipo en San Alfonso												
HIST	47.6	43.1	40.7	37.1	36.0	43.9	66.7	113.5	158.2	145.7	100.3	69.0
A2	44.3	43.4	41.5	38.8	39.5	51.8	69.0	88.7	110.9	102.7	77.0	59.9
B2	45.2	42.7	39.6	34.8	33.8	43.6	61.9	86.1	123.6	110.8	81.8	63.2
Mapocho en Los Almendros												
HIST	2.3	2.8	4.2	4.8	5.0	7.1	10.8	12.3	11.6	7.9	4.7	3.1
A2	2.0	2.7	4.3	5.0	5.8	8.8	10.8	8.9	6.9	4.8	3.3	2.6
B2	2.0	2.7	4.1	3.9	4.3	6.9	9.3	8.6	8.0	5.3	3.5	2.7

Values correspond to cubic metres per second and the 2045–2065 period

There is, nevertheless, a high uncertainty associated with glaciers in Chile's central region. Much has yet to be quantified in terms of total existing glacial mass, and estimates of melt contributions to river systems refined. For further details, the reader is encouraged to review works, for example, by Pellicciotti et al. (2007, 2008) and Petersen and Pellicciotti (2011).

Although vital in many ways, "secondary variables" such as radiation, wind and relative humidity suffer from the absence of long-term records for comparison with climate model outputs. In the Metropolitan Region, these variables have been routinely measured in the past decade for agricultural purposes. Hence no statistical downscaling was attempted in this study. Analysis of the variables was confined to comparing GCM outputs for the historical and future climate scenarios, with results showing that no significant changes are to be expected for the region under future climate conditions. Given the large scale of the GCM modelling grid, however, straddling the MR between ocean- and continental-related cells, further research is needed to refine these variable estimates.

2.5 Conclusions

This chapter summarizes the main results obtained from estimates of future climate and hydrological conditions for the Metropolitan Region of Santiago. Results were presented for the 2045–2065 time period and based on the direct downscaling to local conditions of GCM projections (daily values) measured at selected meteorological stations. Output from multiple GCMs was included in the analysis in order to estimate the uncertainty affecting these projections.

Downscaled climate projections show approximately a 1–2 °C warming for the future period 2045–2065 at most stations in the region. Additionally, days with maximum temperatures above 30 °C increase in the order of 25–45 days per year (A2 scenario), depending on the station under review; this represents approximately a 30 % relative change in the annual number of days with very high temperatures.

Higher temperatures lead to higher elevations in the zero isotherm line, which in turn increase the storm run-off from higher elevation catchments. Minimum temperatures also increase, but unlike maximum temperature results, these increase more significantly at Pirque and El Yeso stations, which are located near or within the mountain region of the area. Precipitation projections are based on two indicators: average monthly values (climatology) and distribution of the number of days with precipitation.

Precipitation amounts may decrease for almost every month. In the worst case, projected reductions in precipitation vary between 10 and 30 % (20–100 mm less precipitation each year, depending on the specific location assessed). In addition, most models project fewer days with precipitation and lower precipitation rates during those days. Due to lower precipitation rates, a general decrease in stream flow magnitude is expected for the Maipo and Mapocho rivers, the most important streams in the MRS. The overall conclusion is that the city of Santiago will be both drier and hotter in the future, with a high number of days of extreme temperatures and increased drought during winter and summer.

The work presented here attempts to organize various data sources coherently into a useful decision-making product. It presents likely changes in climate for a specific time frame in the future and two scenarios that allow for further estimates on climate change related impacts and the subsequent development of specific adaptation measures (cf. Chaps. 4–8 in this volume).

Some limitations persist, however, and should be taken into account in future investigations. On the subject of methodology, only a few stations have a sufficiently long period of record to establish downscaling relations with simulated values, so that spatial interpolation of the data was confined to precipitation and temperature. In the case of precipitation data, orographic effects expressed through an elevation gradient were observed from values compiled from stations in adjacent watersheds, plus the differences observed within the Metropolitan Region. For temperature, no clear spatial pattern was found, and an approximate elevation gradient was adopted in order to estimate future zero isotherm line positions. Regarding GCM data, the current generation of GCMs (IPCC IV) does a good job in capturing the local climatology at the oceanic cells off the coast of Chile, but no cells fall exactly at the location of the MRS. Furthermore, this generation of models does not capture the low frequency climate variations that influence a great deal of the MRS weather, such as ENSO and the Pacific Decadal Oscillation (PDO). It is expected that the next GCM generation (IPCC V) will represent these circulation patterns more successfully; therefore, projections of related phenomena such as maximum daily precipitation and extreme seasonal events (e.g., droughts) should be updated as they become available. The projection of secondary variables such as relative humidity, radiation or wind suffers from a higher level of uncertainty than that of precipitation and temperature, because these variables are much more dependent of local conditions; therefore regional extrapolation from observed data might not represent accurately the predominant conditions at all locations. Glacier characterization is incomplete in much of Chile, and this is also true for the MRS. Thus, stream flow projections should improve (particularly during dry periods) when hydrologic models are able to represent this component more adequately. This research is currently ongoing.

Overall, regional extrapolation of point information is rather uncertain, because the spatial coverage of meteorological stations is insufficient. Compared to developed countries, the MRS has very few data points available, and a large geographical region (the upper Mapocho and Maipo river basins) remains unmonitored. Vertical gradients of precipitation and temperature could and should be verified with dedicated monitoring campaigns, and in general a better understanding of meteorological and hydrological processes in the Andes is required to assess future water resource availability.

Based on the data presented in this chapter, explorative scenarios are designed and presented in Chap. 3, and applied to climate change impact studies in Chaps. 4–7. Hence this chapter forms the basis for identifying, finally, adaptation needs for the MRS. As highlighted earlier, uncertainty is high—as in all climate change projections—and should be considered carefully while working with these results.

References

Borquez, R., Larraín, S., Polanco, R., & Urquidi, J.C. (2006). *Glaciares chilenos*. Reservas estratégicas de agua dulce para la sociedad, los ecosistemas y la economía. Santiago de Chile. Available at: www.sustentarse.cl/publicaciones/glaciares.pdf

Bown, F., Rivera, A., & Acuña, C. (2008). Recent glacier variations at the Aconcagua basin, central Chilean Andes. *Annals of Glaciology, 48*, 2008.

Carrasco, J. F., Casassa, G., & Quintana, J. (2005). Changes of the 0 ° C Isotherm and the Equilibrium Line Altitude in Central Chile During the Last Quarter of the 20th century/ Changements De L'isotherme 0 ° C Et De La Ligne D'équilibre Des Neiges Dans Le Chili Central Durant Le Dernier Quart Du 20ème Siècle. *Hydrological Sciences Journal, 50*(6).

Casassa, G. (1995). Glacier inventory in Chile: Current status and recent glacier variations. *Annals of Glaciology, 21*, 317–322.

Cepal – La Economía del Cambio Climático en Chile. (2009). *Síntesis. Documentos de Proyectos – 288 LC/W.288 Comisión Económica para América Latina y el Caribe.*

Conama – Comisión Nacional Del Medio Ambiente. (2006). *Estudio de la Variabilidad Climática en Chile para el Siglo XXI. Santiago de Chile.*

Cortés, G., Vargas, X., & McPhee, J. (2011). Climatic sensitivity of stream flow timing in the extratropical western Andes Cordillera. *Journal of Hydrology, 405*(1–2), 93–109. doi:10.1016/j.jhydrol.2011.05.013.

Cortés, G., Schaller, S., Rojas, M., Garcia, L., Descalzi, A., Vargas, L., McPhee, J. (2012). *Assessment of the current climate and expected climate changes in the Metropolitan Region of Santiago de Chile*. UFZ-Report 03/2012, Helmholtz Center for Environmental Research UFZ, Leipzig.

Dirección General de Aguas. (2010). *Balance de masa del glaciar Echaurren Norte temporadas 1997–1998 a 2008–2009*. Santiago de Chile.

Falvey, M., & Garreaud, R. D. (2009). Regional cooling in a warming world: Recent temperature trends in the southeast Pacific and along the west coast of subtropical South America (1979–2006). *Journal of Geophysical Research, 114*, D04102. doi:10.1029/2008JD010519.

Garín, C. (1986). *Inventario de Glaciares de los Andes Chilenos desde 18° a los 32° Latitud Sur*. Santiago. Available at: http://bibliotecadigital.ciren.cl/gsdlexterna/collect/bdestudi/index/assoc/HASHb34d.dir/DGA072.pdf

IPCC – Intergovernmental Panel on Climate Change. (2000). *Special report on emissions scenarios (SRES)*. Available at www.ipcc.ch/pdf/special-reports/spm/sres-en.pdf

IPCC – Intergovernmental Panel on Climate Change. (2001). *Climate change 2001*. The scientific basis. IPCC third assessment report. Cambridge University Press, pp. 881. Available at www.grida.no/publications/other/ipcc_tar/

IPCC – Intergovernmental Panel on Climate Change. (2007). *Climate change 2007*. In S. Solomon, D. Qin, M. Manning, Z. Chen, M. Marquis, K. B. Averyt, M. Tignor, & H. L. Miller (Eds.), *The physical science basis*. Contribution of Working Group I to the Fourth Assessment Report of the Intergovernmental Panel on Climate Change (996 pp). Cambridge: Cambridge University Press.

Le Quesne, C., Acuña, C., Boninsegna, J. A., Rivera, A., & Barichivich, J. (2009). Long-term glacier variations in the Central Andes of Argentina and Chile inferred from historical records and tree-ring reconstructed precipitation. *Palaeogeography Palaeoclimatology Palaeoecology, 281*(3–4), 334–344.

Marangunic, C. (1979). *Inventario de glaciares hoya del río Maipo*. Santiago de Chile: Dirección General de Aguas, Ministerio de Obras Públicas. Available online at: http://documentos.dga.cl/GLA1046v5.pdf

Pellicciotti, F., Helbing, J., Araos, J., Corripio, J., Favier, V., & Rivera, A. (2007). *Studying the energy balance and surface melt at the location of an automatic weather station on a glacier of the dry Andes*. Juncal Norte Glacier, Central Chile. Geophysical Research Abstracts 9, 07745.

Pellicciotti, F., Helbing, J., Rivera, A., Favier, V., Corripio, J., Araos, J., et al. (2008). A study of the energy balance and melt regime on Juncal Norte Glacier, semi-arid Andes of Central Chile, using melt models of different complexity. *Hydrological Processes, 22*(19), 3980–3997.

Peña, H., & Nazarala, B. (1987). Snowmelt-run-off simulation model of a central Chile Andean basin with relevant orographic effects. *Large Scale Effects of Seasonal Snow Cover*, 215(166), 161–172 (IAHS Publication)

Petersen, L., & Pellicciotti, F. (2011). Spatial and temporal variability of air temperature on a melting glacier. Atmospheric controls, extrapolation methods and their effect on melt modeling, Juncal Norte Glacier, Chile. *Journal of Geophysical Research, 116*(23), D23109.

Quintana, J., & Aceituno, P. (2006). *Trends and interdecadal variability of rainfall in Chile*. 8 ICSHMO, April, Foz de Iguazu.

Rivera, A., Casassa, G., Acuña, C., & Lange, H. (2000). Variaciones Recientes De Glaciares En Chile. *Revista De Investigaciones Geográficas, 34*, 29–60.

Rivera, A., Acuña, C., Casassa, G., & Bown, F. (2002). Use of remote sensing and field data to estimate the contribution of Chilean glaciers to the sea level rise. *Annals of Glaciology, 34*, 367–372.

Schmidli, J., Frei, C., & Vidale, P. (2006). Downscaling from GCM precipitation. A benchmark for dynamical and statistical downscaling methods. *International Journal of Climatology, 26*, 679–689.

Vera, C., Silvestri, G., Liebmann, B., & González, P. (2006). Climate change scenarios for seasonal precipitation in South America from IPCC-AR4 models. *Geophysical Research Letters, 33*, L13707. doi:10.1029/2006GL025759.

Von Storch, H. (1999). On the use of 'inflation' in statistical downscaling. *Journal of Climate, 12* (12), 3505–3506. doi:10.1175/1520-0442(1999)012<3505:OTUOII>2.0.CO;2.

Widmann, M., Bretherton, C. S., & Salathé, E. P. (2003). Statistical precipitation downscaling over the Northwestern United States Using numerically simulated precipitation as a predictor*. *Journal of Climate, 16*, 799–816. doi:10.1175/1520-0442(2003)016<0799:SPDOTN>2.0.CO;2.

Wilby, R.L., Charles, S.P., Zorita, E., Timbal, B., Whetton, P., & Mearns, L.O. (2004). *Guidelines for use of climate scenarios developed from statistical downscaling methods*. Available at: www.ipcc-data.org/guidelines/dgm_no2_v1_09_2004.pdf

Scenarios for Future Development

3

Jürgen Kopfmüller

Abstract

The analysis and governance of climate change in urban regions, notably in mega-urban agglomerations, faces a dual challenge: firstly, that of dealing with the increasing complexities and dynamics of the different drivers of development, institutions, and actors; secondly, of considering the limited knowledge of both climate change events and their impacts on natural and social systems, particularly at local level. Given that political and societal decisions must be taken under uncertainty conditions, the scenario method plays a major role in providing decision-makers with a basis from which to generate the relevant orientation and action knowledge. Being well-founded as a tool to cope with such complexities and uncertainties, scenarios are applied since long time in several thematic contexts. In this chapter, basic scenario functions, types, challenges and requirements are addressed and pointed out for the specific context of climate change adaptation efforts. The three-step methodological approach and the conceptual and analytical framework applied to the case of the Metropolitan Region of Santiago de Chile (MRS) are described in detail. Finally, selected methodological responses to challenges associated with the analysis and advisory efforts to improve adaptive capacities in strategic urban planning in this regional context are highlighted.

Keywords

Santiago de Chile • Scenarios • Climate adaptation • Strategic urban planning

J. Kopfmüller (✉)
KIT Karlsruhe Institute of Technology, P.O. Box 36 40, 76021 Karlsruhe, Germany
e-mail: juergen.kopfmueller@kit.edu

K. Krellenberg and B. Hansjürgens (eds.), *Climate Adaptation Santiago*,
DOI 10.1007/978-3-642-39103-3_3, © Springer-Verlag Berlin Heidelberg 2014

3.1 Scenarios: Pictures of the Future

Societal development, particularly in mega-urban agglomerations, is characterized by a growing complexity, diversity, interdependencies and dynamics, making predictions about the future more difficult, if not impossible—despite the application of progressively sophisticated forecasting methods. Accordingly, the analysis and management of future development is carried out increasingly under uncertainty conditions. This is especially true in the long-term perspective, a prerequisite for dealing with climate change issues. Nevertheless, looking into and planning the future is an intrinsic aspiration of mankind.

The scenario technique has for many decades been a widely applied and proven tool in futures research and planning to deal adequately with complexities and uncertainties (e.g., Sondeijker et al. 2006; van Notten et al. 2003). It has been used in strategic planning decisions since the 1970s, particularly in international companies.

The growing crisis of traditional forecasting methods—whether quantitative or qualitative, trending, extrapolation or multivariate—in the context of regional economic, demographic, technological or ecological analysis, was the main reason to start using scenarios (Swart et al. 2004; Fink et al. 2004). "Path dependency" is undoubtedly one of the key deficiencies of forecasting, i.e., in a linear and deterministic view of development, the future is seen as a continuation of and conditioned by the past (e.g., Berkhout 2002; Sardar 1999; de Laat 1996). In many cases, this fails to correspond with reality and satisfactorily reflect on new pressures such as climate change or the development of new technologies, or indeed to consider that the past may even be a misleading guide to the future. This clearly indicates the shortcomings of path-dependent thinking.

Scenarios are basically defined as descriptions of possible and plausible future states, including development paths that potentially lead there (Kosow and Gaßner 2008). The core idea of this method is to "design" and analyse alternative future development options rather than to predict the future. In this sense, scenarios are 'conservative' with regard to our ability to 'predict' the future. They are neither correct nor incorrect, but rather representations of possible futures. They provide a basis for discussion, consensus generation, the setting of political and planning objectives and the design of measures to meet these objectives.

The essential functions and aims of working with scenarios are to provide a framework to think about the future in an organized way and create a societally shared vision of potential futures (Volkery and Ribeiro 2009; Kosow and Gaßner 2008; Wiek et al. 2006). Scenarios also have an important communicative function, since they involve the exchange of ideas and perspectives among actors and draw public attention to specific topics. Scenario planning as a systematic and strategic planning method (Lindgren and Bandhold 2009; Schoemaker 1995) can thus be a powerful platform to explore and integrate diverse sources of knowledge and to stimulate imaginative ideas and responses. Consequently, scenarios can contain an early warning function by indicating specific problem phenomena and cause–effect relations, and by creating awareness for a fitting and timely response.

While predictions respond to the question "What *will* happen?" in the future, two ideal type scenario approaches addressing different perspectives can be distinguished (Kosow and Gaßner 2008; Börjeson et al. 2006): (1) explorative (descriptive) scenarios and (2) normative scenarios. Explorative scenarios respond to the question "What *can* happen?" Taking the present as the starting point, they describe possible future events or development paths, regardless of their desirability or probability of occurrence in detail. They consider optional developments of key drivers, possible consequences and societal action, and present a vision of the future with "if-then statements" (*if* certain drivers of the future develop in a certain way, *then* this or that may happen). The main function here is to lay bare unpredictable elements and render key development factors and strategic cause–effect relations transparent. Normative scenarios, in contrast, respond to the question "What *should* happen?". They incorporate values and interests, focusing on desired future situations or objectives—as, for instance, a specific reduction in CO_2 emissions—and on ways of achieving these goals or minimizing the obstacles. This scenario type begins at a certain future point in time and looks back at the present.

The choice of a certain scenario type is primarily determined by the objectives of the according process and its specific context. Although this dichotomous distinction in the scenario approach is helpful, as a model it is far from ideal. Scenario work often requires decisions and selections, and in this sense scenarios will always be normative, at least implicitly. Hence, a combination of both approaches, explorative and normative, has become common practice.

Bearing this in mind, scenarios have the advantage of providing political and societal decision-makers with an enhanced knowledge base to determine development goals and design strategies to meet these goals. This is done by identifying major changes, threats and opportunities that arise in alternative futures, highlighting possible impacts, and drawing up robust measures against long-term uncertainties, i.e., effective and socially acceptable action in possible futures.

Clearly, future change is a product of current decisions. Empirical evidence demonstrates that scientific findings and proposals for strategies based on scenarios undergo scrupulous reflection and are therefore of major relevance to political or civil society addressees, as potential future development alternatives and impact factors are more accurately taken into account. This learning function (e.g., Berkhout et al. 2002) is crucial to suitably advising political decision-makers on a scientific basis, on the one hand, and a precondition for (successful) strategic planning in the context of future uncertainties, on the other. Proper recognition of exogenous economic, social or cultural factors that shape those decision-making processes is vital if simplistic assumptions are to be avoided and efficacious decisions made in the interest of maximizing well-being into the future and minimizing risks.

This chapter gives an overview of the scenario approach designed to estimate selected impacts of climate changes in the MRS and to develop appropriate adaptation measures as presented in Chaps. 4, 5, 6, 7, and 8. Section 3.2 below outlines the previous use of scenarios in climate change related analysis in general, and the according procedure of the IPCC in particular. Section 3.3 describes the scenario approach designed and applied for the Santiago case. Finally, Sect. 3.4 concludes with some general and case-specific reflections on the challenges of using scenarios in the climate context and ways of dealing with them.

3.2 Scenario Use in Climate Change Analysis

Climate change concerns the medium and long-term perspective in terms of temperature, precipitation and other elements of the climate system, and of its impact on societal sectors and institutions. Thus, tackling these changes demands future-oriented planning, which in turn requires current decisions to initiate and shape a suitable transition management of organizational and societal systems. Since the Fourth IPCC Assessment Report (2007) at the latest, there has been a consensual view that for a suitable climate policy, adaptation to expected and potentially unavoidable climate changes is as important as the mitigation approach hitherto in the scientific and political focus (Betsill and Bulkeley 2007). Although both approaches, mitigation and adaptation, are interrelated to a certain extent (cf. Chap. 1), it should be noted that mitigation policies are implemented for the most part at national and international level. Adaptation activities, on the other hand, must be carried out at local and regional level, and their benefits are mainly local to regional in scale, depending on how substantial climate change impacts are on the respective geographic locations and other locally contextualized factors (IPCC 2007).

As climate change occurs in a highly complex structure of interacting social, economic, political and environmental drivers, institutions and technological factors, this calls for new advanced governance approaches and knowledge bases, posing new scientific challenges. Adaptation, seen as a dynamic social and institutional process, is linked to strong, albeit specific, uncertainties similar to those in the mitigation approach: since future changes in climate variables are highly uncertain in terms of timing and magnitude of events and their effects, particularly at regional level, this applies all the more to potential climate change impacts on social sectors, institutions or technologies that could become the subject of adaptation considerations. Furthermore, adaptation activities have up to now been less sound in theoretical and conceptual terms than mitigation measures. The validity, operability and effectiveness of adaptation measures is both uncertain and controversial, primarily due to conflict with other goals, their social and distributional effect on socio-economic groups, and potential obstacles to implementation. This is a result of the current lack of figures based on experience compared to those available for mitigation measures. Finally, participation is crucial, since adaptation strategies frequently affect people's daily lives at the local level.

Against this background, the basic socio-political objective (and challenge) is to strengthen the adaptive capacity of society and its ability to manage natural and social resilience so as to reduce vulnerabilities to climate changes (IPCC 2007; Gallopin 2006). This requires a framework of adaptive governance or transition management (e.g., Jacobson et al. 2009; Brunner and Steelman 2005; Folke et al. 2005) that will govern climate adaptation processes at the science–policy–society interface, integrating various scientific disciplines, societal sectors, spatial and political levels and scales, and stakeholders. Here, traditional linear and sectorally organized planning and decision-making methods, based for the most part on expert knowledge and analysis, are unlikely to be effective. Policy-makers, decision-makers and practitioners dealing with climate change adaptation are making more and more use of scenario planning methods. The results allow for planning and management to consider a variety of potential outcomes and to respond in a more adaptive, long-term, flexible and effective way compared to the predominantly linear, one-dimensional planning currently in operation (e.g., Biggs et al. 2011; Jones and Preston 2010).

From the outset, the IPCC has employed scenarios to deal with future data based on historic facts and figures and to suggest alternatives to the way in which climate change impacts might develop in the future. These scenarios shape the panel's analysis of opportunities for a range of climate change related mitigation and adaptation activities and their consequences. Four scenarios entitled A1, A2, B1, B2 were created, based on storylines that can be briefly described as rapid economic and more efficient technologies (A1), self-reliance and preservation of local identities accompanied by slower growth and technological change rates (A2), more service- and information-oriented economies, material intensity reductions and cleaner technologies (B1), and local solutions to sustainability based on more diverse technologies and social equity (B2). For all four scenarios, common assumptions on global population and GDP growth were adopted to achieve a certain degree of harmony in the framework data. For the approach presented in this book, the scenario alternatives A2 and B1 were used as a starting point for down-scaling global climate change model results to the regional level of the MRS (for details, Chap. 2 in this book). In recent years, these two scenario options have been the preferred description of future changes in temperature and precipitation, and regarded as the most probable and the most manageable by policy-makers.

Since the climate change debate and the work of the IPCC have emphasized the need for a paradigm shift in the global development model, away from carbon dependency and towards alternative models of development, life styles and quality of life, the implications must be reflected at multiple scales.

Scenarios are statements about the future rather than projections and should therefore not be expected to represent realities. Instead, they are to be understood as possible stories of potential futures around which the future can be discussed from a common core. Complexities, uncertainties, and thus scenario-based thinking gain even greater relevance if the more decarbonized and dematerialized development model to cope with climate change goes hand in hand with sustainable development. This will lead to expansion of the climate-inherent, long-term perspective to reflection on social change and justice considerations. Since climate change modelling points to the end of the century, local and regional development models and institutional planning models need to do the same.

3.3 The Applied Scenario Approach

The design and analysis of the scenarios in this book was carried out in a three-step approach. This included (1) the development of framework scenarios for the case of Santiago de Chile and Chile, taking into account the principal driving factors of global development as well as qualitative and quantitative information, (2) the translation of these scenarios into the context of energy, land use, water and vulnerability by estimating potential climate change impacts on each of these fields, and (3) the analysis of possible adaptation measures and the relevant interdependencies as a response to the expected impacts, and finally the proposal of a suitable mix of measures. The results of steps (2) and (3) are presented in Chaps. 4, 5, 6, 7, and 8 of this book.

In step 1, two explorative framework scenarios entitled "Business as Usual" (BAU) and "Collective Responsibility" (CR), written in the form of storylines or narratives, were developed for the MRS for the period up to 2050, with reference to existing global scenario studies. Elaborated by international organizations such as Shell, the United Nations Environment Programme (UNEP) and the US National Intelligence Council or by international futurology institutes, these studies constituted the basis for selecting the driving factors that most affect societal development, such as economic, demographic or governance-related developments, changing consumption patterns, or technological progress, and for describing plausible options as to how these factors might develop in the coming decades (Woll 2010). The fundamental ideas ("philosophies") contained in the BAU and CR development paths are listed in Box 1.

Box 1: Basic "philosophies" of the two MRS scenarios

Business as Usual

- Emphasis on market and deregulation in production and consumption of goods and services, with strong private sector influence in decision-making with regard to policy, planning and investments. The public sector is subsidiary, confining its role to market regulation as opposed to strategic planning.
- Continued trade liberalization based on comparative advantages; the export sector remains the driver of the national economy, based on minerals, forestry, fisheries and agriculture. The MRS development is focused on the services associated with these natural resource export sectors.
- The MRS continues to be the hub of national politics and economic development, concentrating wealth generated in other regions. It continues to attract national and international migrants, leading to increasing demands for goods and services, land and transport. Water and energy costs continue to rise, not least due to climate change impacts.

(continued)

Box 1 (continued)

- Urban sprawl persists alongside larger plot sizes and smaller households. Investments in infrastructure and transport increase to accommodate this morphology. Traffic congestion increases, public and private transport infrastructures increase in parallel.
- Civil society rises in scale and influence, and has a diverse agenda. Sustainable development and climate change concerns gain in significance among these groups. General awareness of climate change concerns nevertheless remains low and is reduced to a strong cost-benefit logic, with little media coverage.
- Income and service provision inequalities persist with little change; individual mobility of different socio-economic groups is also constrained. Crime and violence continue to dominate public perceptions and concerns. Individual life-style development is characterized by an increasing focus on material compared to non-material consumption patterns.

Collective Responsibility

- The state plays a lead role in decision-making, heading towards more strategic planning. Markets are more regulated, particularly for basic services and transport provision. In policy and evaluation of measures, social equality is prioritized over economic growth. Decision-making involves high levels of participation (social capital building) with public, private and civil society actors.
- Subsidiary influences public policy, with intra-national and intra-MRS decentralization towards a multi-polar region of decision-making and public and private investments. Sustainable development and climate change measures are considered explicitly in these activities.
- An export-oriented development model remains most influential, but considerable investment in education and research slowly changes this profile, aiming at more diversification of the economy and innovation activities.
- This new development model has a strong decentralization component, built on regional centres apart from the MRS, which focuses on higher education and tourism. This reduces migration to the MRS and induces firm migration from the MRS to other regions.
- Chile plays a more influential role in innovative climate change adaptation strategies in Latin America. The MRS becomes a centre of international climate change leadership. Climate change issues are built into all multilateral and bilateral initiatives.
- Service provision, including energy, is shaped by strategic planning objectives for socioeconomic justice in adaptation. Substitution of traditional energy sources and eco-efficiency are central to this process.

(continued)

Box 1 (continued)

Transport and infrastructure in general are more oriented towards public provision. Urban sprawl is discouraged and compaction is favoured in Santiago, while other poles in the region are encouraged to grow, forming a functional MRS urban network.

- Education, in particular life-long learning, becomes central to adapting society and the economy to twenty-first century challenges such as climate change. Crime rates fall and employment opportunities rise.
- Climate change is widely discussed in the media, progressively beyond the narrow cost-benefit logic, with local commitments to adaptation endeavours. Efforts to increase local resilience rise. Non-material consumption is emphasized as distinct from material consumption, with a rise in diverse life styles and livelihoods.

On the whole, urban development in the MRS, i.e., transport infrastructure, new buildings as of 2005, will dominate urban life and urban development for many decades, determining opportunities and reducing options for urban design and planning into the future. Additionally, the above-mentioned driving factors, the current situation in the MRS and the two possible future development paths were affected by two external shocks: (1) the financial crisis 2008–2010 and its impact on economic and social development, and (2) the earthquake in the Maule region in February 2010. The financial crisis reduced GDP growth temporarily due to reduced exports, a decline in loans from the financial sector and a drop in domestic consumption and investments. To alleviate the tremendous impact of the earthquake on buildings and infrastructure, public and private reconstruction programmes were implemented. This led to a growth in the national GDP, whereas the GDP in the MRS was reduced—as a net effect of shifting investments from the MRS to the affected region and a surge in the demand for MRS companies to produce goods and services required in the Maule region (e.g., Magud 2010).

The scenario framework furthermore built on quantitative estimates for the future development of several variables that characterize and heavily influence the development of the MRS, such as GDP growth rate, population growth, number of households and economic structure. Exemplary, the data for three selected variables and brief explanations are presented in Table 3.1.

These estimates are based on official data, literature studies, expert knowledge and discussions with stakeholders. They mark important boundary conditions for the estimation and assessment of climate change impacts on the region and for the development of proposals for suitable adaptation measures.

In the second step, two alternative framework scenarios consisting of storylines and data sets were translated into the context of the four thematic fields analysed in detail: the energy sector, the water sector, land use and vulnerability (Chaps. 4, 5, 6, and 7). Contextualized storylines for the two scenario options were elaborated for each thematic field (Chaps. 4, 5, 6, and 7), taking into account expected changes in

Table 3.1 Quantitative variables: trends and scenario estimates for the MRS (2010–2050) (*Source*: Barton et al. (2011))

Variables	History	BAU	CR
Total population (52 municipalities) (in millions)	1990: 5.19 1995: 5.70 2000: 6.17 2005: 6.54	2010: 6.88 2020: 7.46 2030: 8.00 2040: 8.30 2050: 8.50	2010: 6.88 2020: 7.45 2030: 7.60 2040: 7.80 2050: 7.90
Explanations	INE data	Official INE projection to 2020. The projection to 2030 follows the same tendency. From 2030 on, growth rates further decrease according to the national projection of INE up to 2050. A demographic transition phase (peak) is assumed for 2050, similar to the global trend.	Lower growth rates due to effective decentralization strategies at the national level. Less in-migration. Trend continued to 2050. Higher growth occurs in intermediate cities, particularly in the southern regions where water and energy resources are more accessible.
GDP yearly growth rate (in %) real (i.e., excluding inflation)	2009: 1,2 *Source*: Banco Central de Chile (2011a, b, c, d)	2010: 4,0 2020: 5 2030: 4 2040: 2,5 2050: 2,5	2010: 4,0 2020: 4,5 2030: 3,5 2040: 2 2050: 2
Explanations		Negative growth in 2009 due to the crisis, but less so than at national level. According to the increasing percentage of national GDP, rates rise compared to the national level. Considering limits to growth also affects the MRS, although later than in CR.	Decentralization and falling contribution to national GDP leads to lower rates compared to BAU and the national level. This is also due to substantial limits to growth considerations beyond 2020 following the currently emerging international debates.
Persons per household	1992: 3.8 2002: 3.7 2006: 3.7 <u>for comparison</u>: EU lowest: Germany 2010: 2,1 2030: 1,9 EU highest: Estonia, Italy 2010: 3,0	2010: 3.6 2020: 3.3 2030: 3.0 2040: 2,7 2050: 2,5	2010: 3.6 2020: 3.4 2030: 3.2 2040: 2,9 2050: 2,7
Explanations	INE (1992, 2010), Casen (2006)	Changing life styles and demographics will lead to smaller households, following the MRS trend since 1992 and the trend in	Social policy and community programmes generate stronger family cohesion in diverse household types and lead to less uni-personal

(continued)

Table 3.1 (continued)

Variables	History	BAU	CR
		nearly every industrialized country since the 1980s. Less overcrowding due to more housing also contributes to this trend. It is assumed that in 2050 the MRS will reach the 2010 EU average.	dwellings for social and environmental reasons. This leads to a reduction in the decreasing numbers of persons per household compared to BAU. The trend in relation to BAU remains up to 2050. Densification of the existing urban core is more likely given the introduction of smaller housing units for smaller households. This also occurs in other MRS regional growth poles. Large house-garden units decline relative to the total housing stock.

the main climate variables, on the one hand (Chap. 2), and previous trends and future estimations for the most relevant driving factors used in the framework scenarios, on the other. On this basis, it was estimated how these driving factors might affect the thematic fields. Here, questions such as how changing temperatures or precipitation might affect energy generation and consumption, water supply and demand or land-use patterns had to be answered, and how alternative demographic, social, economic, technological, life-style or governance related development paths might influence energy supply and demand, the water sector, and mobility or settlement patterns over the next 40 years. Summarized outlines of these contextualized storylines are presented in Chaps. 4, 5, 6, and 7 of this book.

Analytically, a key challenge was to deal appropriately with the uncertainties and margins of the regional climate change estimates (Chap. 2), which existed even according to the IPCC scenario assumptions. A number of approaches were possible here: (1) the use of average numbers based on margins of change provided by the climate models in both scenarios, (2) the use of margins of change to carry out a sensitivity analysis within these margins for both scenarios, or (3) to link each of the two framework scenarios to one of the IPCC scenarios and the particular climate changes resulting from them. A combination of option 1 and option 3 was applied to estimate the different climate change impacts on the respective thematic fields (Chaps. 4, 5, 6, and 7).

Tackling uncertainties valid for all estimates, including quantitative values in the framework scenarios and estimated climate change impacts, is a crucial element of scenario analysis. Similarly vital to the scenario-based process is, thus, the transparency of assumptions used and the open discussion of assumptions and preliminary results with the stakeholders involved.

The third and final step comprised several elements. The most acute impacts of estimated climate changes on the different thematic fields were first of all identified in order to prioritize response strategies. Based on expert knowledge and a review of the literature, possible field-specific adaptation measures were selected and evaluated. Preliminary results were discussed extensively with the stakeholders and finally a set of suitable measures ranked according to their effectiveness and their embeddedness in existing institutional structures presented (Chap. 9). Due to the specific situation in the energy sector, where the distinction between adaptation and mitigation measures is particularly difficult, stronger emphasis was placed here on measures that can clearly be described as adaptation measures, and, beyond that, on the very close link between mitigation and adaptation approaches in this field. Finally, in order to provide input for an appropriate mix of measures and to avoid mutual counter-productivities, possible interdependencies of and linkages between selected measures were described for their potential to enhance or reduce intended positive effects.

3.4 Challenges and Responses

The scenario working steps outlined in this chapter marked both the core and the thread of the analysis, and, as expected, were accompanied by several challenges. In general, estimates of future development options are based on suitable information concerning previous and current trends, as well as on normative views of what should happen. Hence, the crucial challenge is to combine the necessary normative elements and comply with basic scientific standards in order to provide comprehensible orientation knowledge based on scenario analysis.

Developing the framework scenarios (step 1) posed a challenge, since state-of-the-art knowledge on global scenario studies had to be worked out and estimates made of the possible impacts of global driving factor trends on the situation in Chile and the MRS, and of how these could be translated into the future development analysis of the region. This, in turn, required suitable "local" knowledge of the relevant contexts, the validation of this knowledge and the involvement of local experts and stakeholders concerned. Such knowledge was of particular relevance when it came to projections for the quantitative variables and the consideration of remaining uncertainties—bearing in mind that the slightest variation in variable values can affect results substantially.

Step 2 required estimates of the impacts of these variables on regional thematic field development, creativity in designing field-specific narratives that included visionary and scientifically valid elements, local and field-specific knowledge.

Finally, step 3 called for the pre-selection, discussion, analysis and ultimate choice of appropriate adaptation measures for the individual fields and with respect to a pertinent combination of measures. Selection criteria had to be justified, and quantitative and qualitative analytical methods chosen and applied.

To deal aptly with these challenges, apart from comprehensive reviews of the literature and of scientific, political and societal debates, three elements were consistently applied in the analysis described in this chapter:

- Firstly, close cooperation from the outset between the involved scientists from different disciplines and cultures on the conceptual work approach, methodologies, and decisions about the involvement of experts/knowledge.
- Secondly, transparency of assumptions and selections made in the course of the working steps, of methodological deficiencies and problems, and how they were handled, and of the remaining uncertainties in the results.
- Finally, the systematic involvement of the relevant stakeholders in each of the working steps.

Implementing approaches that entail interdisciplinary and "inter-cultural" co-operation, transparency and stakeholder participation is, particularly in the case of scenario analysis, a precondition for process addressees and interested communities to gain a better understanding of scenario-based scientific findings and their reproduction. Empirically, this procedure increases acceptability significantly and the factual acceptance of using these results as orientation and action knowledge for decision-makers—in the case described here, of their integration into upcoming climate change related strategies and plans in the MRS. In this science-policy interface of communication and cooperation, normative elements are inevitable. Rendering them transparent and integrating them into analysis is therefore essential. Scenarios and their incorporated thinking in plausible if-then-options are the best frame and tool to achieve this.

The ultimate goal of the scenario approach is to generate consensus about the desirability, or not, of certain future paths and their operability, and to identify the chief obstacles to implementation of the intended strategies (e.g., Moser and Ekstrom 2010). This in turn prompts discussion on the types of measures best suited to or that modify existing context conditions—here for the case of climate change adaptation. All of these steps are vital to effective planning. Since climate change adaptation requires actions today and throughout the coming century, the sooner inter-generational, long-term thinking and planning is incorporated into decision-making processes, the more resilient to climate changes we can expect city regions to become.

References

Banco Central de Chile. (2011a). *Indicadores Economicos y Sociales de Chile 1960-2000*. Santiago de Chile. (http://www.bcentral.cl/publicaciones/estadisticas/informacion/)

Banco Central de Chile. (2011b). *Producto Interno Bruto Regional 2003-2006, base 2003*. Santiago de Chile (http://www.bcentral.cl/publicaciones/estadisticas/actividad-economica-gasto/aeg07b.htm)

Banco Central de Chile. (2011c). *Cuentas Nacionales de Chile 1996-2005*. Santiago de Chile (http://www.bcentral.cl/publicaciones/estadisticas/actividad-economica-gasto/pdf/ctasnac2005.pdf)

Banco Central de Chile. (2011d). *Producto interno bruto por actividad y por region*. Santiago de Chile (http://www.bcentral.cl/estadisticas-economicas/series-indicadores/index.htm)

Barton, J., Kopfmüller, J., & Stelzer, V. (2011). *Project "ClimateAdaptationSantiago" (CAS): Quantitative variables: tendencies and scenario estimations 2010–2050*. Unpublished working paper.

Berkhout, F. (2002). Technological regimes, path dependency and the environment. *Global Environmental Change, 12*(2), 1–4.

Berkhout, F., Herton, J., & Jordon, A. (2002). Socio-economic futures in climate change impact assessment: Using scenarios as 'learning machines'. *Global Environmental Change, 12*(2), 83–95.

Betsill, M., & Bulkeley, H. (2007). Looking back and thinking ahead: A decade of cities and climate change research. *Local Environment, 12*(5), 447–456.

Biggs, C., Edwards, T., Rickards, L., & Wiseman, J. (2011). *Scenario planning for climate adaptation*. Victoria: Victorian Centre for Climate Change Adaptation Research.

Börjeson, L., Höjer, M., Dreborg K., Ekvall, T., & Finnveden, G. (2006). Scenario types and techniques: Towards a user's guide. *Futures, 38*(7), 723–739.

Brunner, R., & Steelman, T. (2005). Towards adaptive governance. In R. Brunner, T. Steelman, L. Coe-Juell, C. Cromley, C. Edwards, & D. Tucker (Eds.), *Adaptive governance: Integrating science, policy and decision making* (pp. 268–303). New York, NY: Columbia University Press.

de Laat, B. (1996). *Scripts for the future: Technology foresight, strategic evaluation and socio-technical networks. The confrontation of script-based scenarios*. Amsterdam: University of Amsterdam.

Fink, A., Siebe, A., & Kuhle, J. (2004). How scenarios support strategic early warning processes. *Foresight, 6*(3), 173–185.

Folke, C., Hahn, T., Olsson, P., & Norberg, J. (2005). Adaptive governance of social-ecological systems. *Annual Review of Environmental Resources, 30*, 411–473.

Gallopin, G. (2006). Linkages between vulnerability, resilience, and adaptive capacity. *Global Environmental Change, 16*(3), 293–303.

INE. (1992, 2002). *Censo de Población y Vivienda*. Santiago de Chile.

INE – Instituto National de Estadísticas. (2010). *Proyecciones de población 1990-2020, por región y comuna*. Santiago de Chile (http://www.ine.cl/canales/chile_estadistico/demografia_y_vitales/proyecciones/DatCom/SalComUsuarios-13Tok.xls)

IPCC – Intergovernmental Panel on Climate Change. (2007). *Climate change 2007: Impacts, adaptation and vulnerability*. Working Group II Contribution to the IPCC Fourth Assessment Report. Summary for Policy Makers. Geneva.

Jacobson, C., Hughey, K., Allen, W., Rixecker, S., & Carter, R. (2009). Toward more reflexive use of adaptive management. *Society and Natural Resources, 22*(5), 484–495.

Jones, R., & Preston, B. (2010). *Adaptation and risk management*. Climate Change Working Paper No. 15. Melbourne: Centre for Strategic Economic Studies, Victoria University Melbourne.

Kosow, H., & Gaßner, R. (2008). *Methods of future and scenario analysis. Overview, assessment and selection criteria* (Studies No. 39). Bonn: German Institute for Development Policy.

Lindgren, M., & Bandhold, H. (2009). *Scenario planning: The link between future and strategy*. London: Palgrave McMillan.

Magud, N. (2010, September 29). *Chile: Strong recovery after devastating earthquake*. In: International Monetary Fund Survey Magazine: Countries & Regions, Washington, DC.

MIDEPLAN – Ministerio de Planificación y Cooperación. (2006). *Encuesta Casen, different years*. Santiago de Chile (http://www.mideplan.cl/casen/)

Moser, S., & Ekstrom, J. (2010). A framework to diagnose barriers to climate change adaptation. *Proceedings of the National Academy of Sciences, 107*, 22026–22031.

Sardar, Z. (1999). *Rescuing all our futures: The future of future studies*. Westport: Praeger.

Schoemaker, P. (1995). Scenario planning: A tool for strategic thinking. *Sloan Management Review, 36*(2), 25–40.

Sondeijker, S., Geurts, J., Rotmans, J., & Tukker, A. (2006). Imagining sustainability: The added value of transition scenarios in transition management. *Foresight, 8*(1), 15–30.

Swart, R., Raskin, P., & Robinson, J. (2004). The problem of the future: Sustainability science and scenario analysis. *Global Environmental Change, 14*(2), 137–146.

van Notten, P., Rotmans, J., van Asselt, M., & Rothman, D. (2003). An updated scenario typology. *Futures, 35*(5), 423–443.

Volkery, A., & Ribeiro, T. (2009). Scenario planning in public policy: Understanding use, impacts and the role of institutional context factors. *Technological Forecast and Social Change, 76*(9), 1198–1207.

Wiek, A., Binder, C., & Scholz, R. (2006). Functions of scenarios in transition processes. *Futures, 38*(7), 740–766.

Woll, T. (2010). *Synopsis of global scenario surveys.* Risk Habitat Megacity project working paper, Karlsruhe.

Part III

Climate Change Impacts on the Urban-Regional Level of Santiago de Chile: Key Sectors and Vulnerabilities

Climate Change Impacts on the Water Sector

4

Helmut Lehn, Laura Margarete Simon, and Melanie Oertel

Abstract

The regional impacts of global climate and socio-economic change will heavily influence the future balance of water availability (supply side) and water demand in the Metropolitan Region of Santiago de Chile (MRS). Reduced run-off in the Maipo-Mapocho river catchment coupled with natural precipitation variability will pose a major challenge for water resource management in the coming years. This chapter elaborates on an impact assessment for the year 2050, which combines two climate scenarios for the supply side with two explorative socio-economic scenarios for the demand side. While adaptive measures for water supplies also involve increasing water storage or recycling grey water in the urban area, adaptive options for water demand focus on upgrading water efficiency in both agriculture and private domestic households. In addition to technical aspects, institutional/policy-based matters and capacity development measures are considered.

Keywords

Climate change • Regional impacts • Water supply • Water demand • Scenarios • Adaptation • Santiago de Chile

H. Lehn (✉)
KIT Karlsruhe Institute of Technology, ITAS-Institute for Technology Assessment and Systems Analysis, Karlstrasse 11, 76133 Karlsruhe, Germany
e-mail: helmut.lehn@kit.edu

L.M. Simon
Cologne University of Applied Sciences Institute for Technology and Resources Management in the Tropics and Subtropics (ITT), Betzdorfer Straße 2, 50679 Köln, Germany

M. Oertel
Pontificia Universidad Católica de Chile Centro Cambio Global Vicuña Mackenna, 4860 Macul, Santiago, Chile

K. Krellenberg and B. Hansjürgens (eds.), *Climate Adaptation Santiago*,
DOI 10.1007/978-3-642-39103-3_4, © Springer-Verlag Berlin Heidelberg 2014

4.1 Introduction

This chapter assesses predicted climate change impacts on the water sector and the attendant societal changes in the Metropolitan Region of Santiago de Chile (MRS). The goal is to design measures to adapt to climate change and to propose shifts in socio-economic parameters until the middle of this century. Following a description of the Maipo-Mapocho water catchment (Sect. 4.2), the status quo and current balance of water availability and water demand will be presented (Sect. 4.3). Based on a description of the applied scenario methodology, future changes in water availability induced by the adverse effects of climate change are assessed. Future water demand patterns are described with two socio-economic scenarios, and a possible future water balance is introduced (Sect. 4.4). Against this background, the need for adaptation to climate change effects in the water sector is derived and possible adaptive measures are presented (Sect. 4.5). The paper closes with some conclusions (Sect. 4.6).

The assessment of climate change impacts faces uncertainties; this study is therefore limited by a number of constraints, which have already been mentioned in Chap. 3 of this book. Uncertainties appear at different levels:

1. data restrictions regarding hydrology and future climate change impacts,
2. unpredictable socio-economic developments, e.g., population growth, economic growth and urbanization patterns, and
3. changes in the infrastructure, e.g., ageing or deterioration of pipe systems, leading to leakage rates in the water supply or sewerage systems.

4.2 The Maipo-Mapocho Water Catchment

The basin of the Maipo river and its main tributary, the Mapocho, lies in Chile's central valley at an approximate mean elevation of 500 m above sea level, surrounded by two Andean cordilleras on the north-western, south-western and eastern side. The catchment area largely corresponds to the area of the MRS, with only a small area belonging to the V and VI region (INE 2008). The Maipo and Mapocho rivers both originate in the glacier zones of the Andes and flow across the MRS (see Fig. 4.1). The Maipo is about 250 km long and its watershed covers approx. 15,400 km^2. The Mapocho is 120 km long, its watershed is notably smaller at about 4,100 km^2 (CNR 2007). The rivers supply 70 % of the potable water demand and around 90 % of the demand for agricultural irrigation in the MRS. In this chapter, water availability and water demand will be analysed for the area of the MRS outside the high Andes, in the west of the Andean cordillera. In the following this area will be referred to as the Santiago Basin. For practical reasons the extent of the Santiago Basin is calculated as the area of the MRS lower than 900 m above sea level, corresponding to approximately 7,000 km^2.

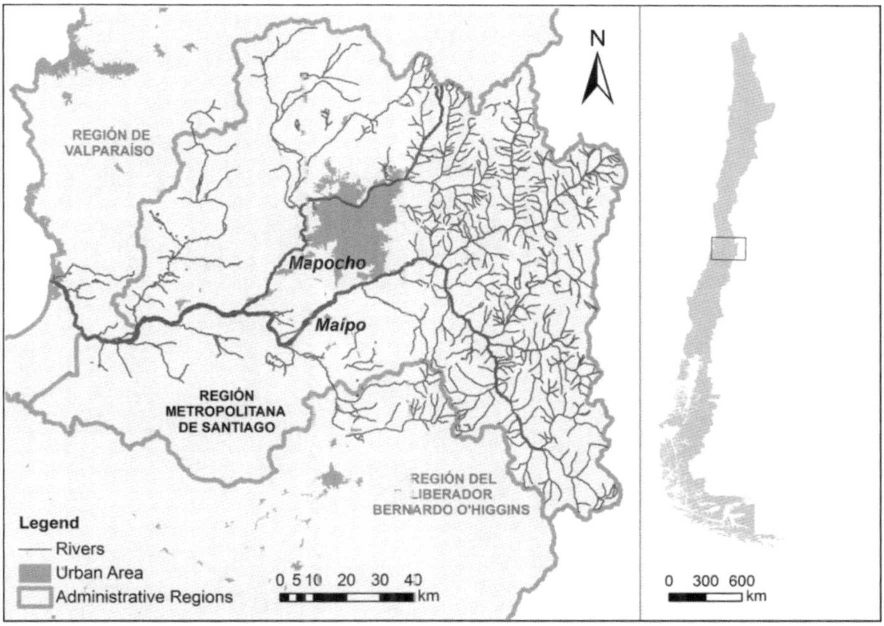

Fig. 4.1 Maipo-Mapocho water catchment (*Source*: SIIT 2013, Cartography: Kiemle)

4.3 Current Status of Water Availability and Water Demand

4.3.1 Current Water Availability

The renewable water resources available in the Santiago Basin are determined by (1) precipitation in the Santiago Basin and (2) the inflow of surface water from the Andes via the Maipo and Mapocho rivers. The temporal availability of water resources in the plains of the MRS is determined mainly by temperature and precipitation, as well as the water storage capacity in the mountainous regions (snow and ice). Precipitation in the MRS varies considerably throughout the year and from one year to another (see Fig. 4.2). This fluctuation is to some extent associated with the El Niño-Southern Oscillation (ENSO). Abundant precipitation can be observed in the course of El Niño events, while precipitation during La Niña events is low (Meza 2005). In the period from 1950 to 2004, the driest year (1968) in Santiago showed precipitation values of 70 mm (Quinta Normal meteorological station) and the wettest year (1987) 713 mm (Lehn et al. 2012).

This inter-annual variation interferes with the intra-annual variation (see Fig. 4.3, left side), leading to highly variable monthly precipitation amounts. The two main water supply sources in the Santiago Basin—precipitation and inflow of surface waters from the Andes Mountains—complement each another (see Fig 4.3).

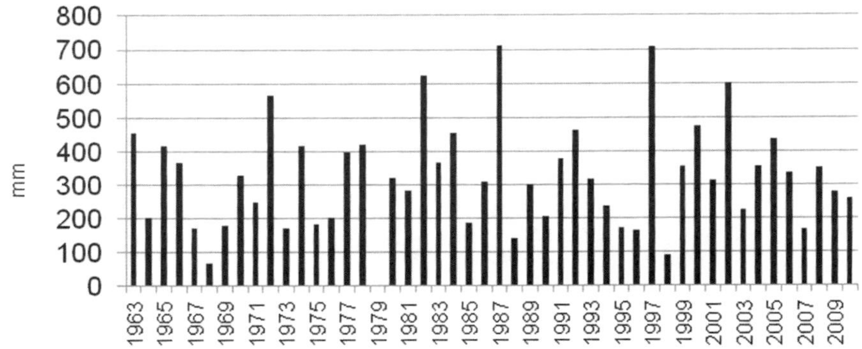

Fig. 4.2 Variations in total annual precipitation in Santiago over time—Quinta Normal station (*Source*: DMC 2012, modified by authors)

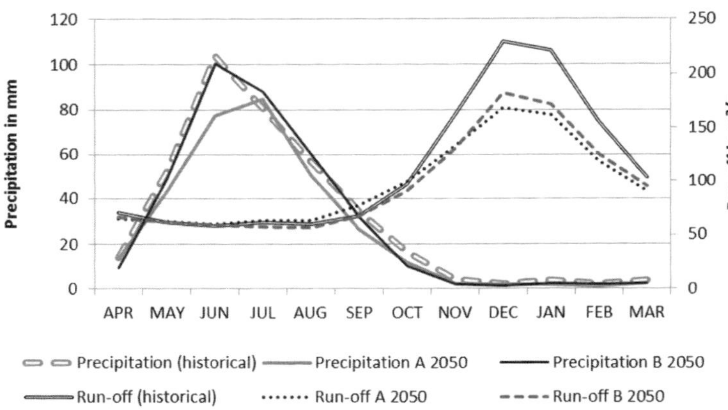

Fig. 4.3 Average precipitation rates from three stations (*left*) and run-off distribution from the Maipo River station at El Manzano (*right*) during the year (historical and A and B 2050 values) in the Santiago Basin (*Source*: Cortés et al. 2012, modified by authors)

When the winter rain ceases, rising spring temperatures induce ice and snow melt in the Andes, and inflowing surface waters take over the water supply in spring, summer and autumn (Kopfmüller et al. 2009).

In general, annual precipitation increases with ascending elevation. As a result of low average annual precipitation values (480 mm) and the fact that almost all of the rainfall is concentrated in the winter months between April and September (Cortés et al. 2012), most water-related needs in the Santiago Basin in spring, summer and autumn depend heavily on melt waters from glaciers and snowfields in the High Andes. In addition to the latter, important water reservoirs are the natural lagoon 'Laguna Negra' and the 'El Yeso' dam, both located in the Andes.

Due to the high inter-annual variability of rainfall in the MRS, the annual mean precipitation and run-off of inflowing rivers is subject to substantial variation. The renewable water resource from precipitation in the Santiago Basin varies between 0.5 and 5 km^3/a. The run-off of the Maipo river varies in the range from 60 to 80 m^3/s in drier and 150 m^3/s in wetter years, averaging at 110 m^3/s (1951–2005).[1] The corresponding data for the Mapocho river is 2–3.5 m^3/s in drier and 9–14 m^3/s in wetter years, averaging at 6 m^3/s (Bartosch 2007). Calculations based on this data reveal that the total annual available amount of renewable fresh water (availability) in the MRS varies between 3 km^3 and 10 km^3, with an average of 6 km^3 (see Table 4.1).

4.3.2 Current Water Demand

The current water demand in the MRS is primarily driven by agricultural needs. In 2007, this sector accounted for approx. 74 % of the total water demand in the region. The share of drinking water of the total annual water demand is about 18 % and for industrial purposes about 8 % (see Fig. 4.4).

According to the latest agricultural census, the MRS accounts for a total of approx. 1,130,000 ha of classified agricultural land, with 136,000 ha (13 %) irrigated by systems such as gravity, sprinkler and micro-irrigation systems (INE 2007; PUC 2011). Irrigation is clearly a water-intensive activity, since this relatively small area accounts for more than two-thirds of the entire water demand in the region. More than 90 % of the irrigated area depends on water withdrawals from surface flows. The total annual agricultural irrigation volume is calculated at 3.2 km^3 (PUC 2011).

The largest water and sanitation company in the MRS is the privately owned Aguas Andinas S.A. (public limited company), which is majority owned by AGBAR (Aguas de Barcelona). As a result of buying up three smaller companies (among them Aguas de Manquehe and Aguas Cordilleras), Aguas Andinas, which is controlled by international corporate groups, dominates the market. Aguas Manquehe and Aguas Cordillera still operate under their former names. The company provides fresh water for over 1,500,000 households in the MRS. In addition, the municipal enterprise SMAPA serves the south-western municipality of Maipú with approx. 180,000 households (SISS 2011). From the data in Table 4.2, an average per capita consumption in the MRS of 79.2 m^3 per year or 217 l per day can be calculated. However, there is a vast difference in consumption between more prosperous households in the eastern parts of the city and low-income households in the more western parts. Areas in the east of Santiago served by Aguas Manquehue reach a consumption level of about 805 l per capita and day, mostly due to irrigation of private green areas and the filling in of swimming pools, while the demand for areas served by the public water and sanitation company SMAPA amounts to roughly 180 l per capita and day (SISS 2011). Water losses in the distribution systems vary

[1] Related to the precipitation rates at Quinta Normal station. Years with precipitation values between 200 and 400 mm are regarded as 'normal years', those above 400 mm as 'wet years', and below 200 mm as 'dry years'.

Table 4.1 Water availability in the Santiago Basin (number of inhabitants: 6.88 million, 2010) (*Source*: modified after Bartosch 2007, data base DGA 2007a)

	Annual renewable fresh water (km^3/a)	Specific available fresh water resource per capita (m^3/cap.*a)
Average	6.2	901
Max (wet) years	10.2	1,483
Min (dry) years	2.8	407

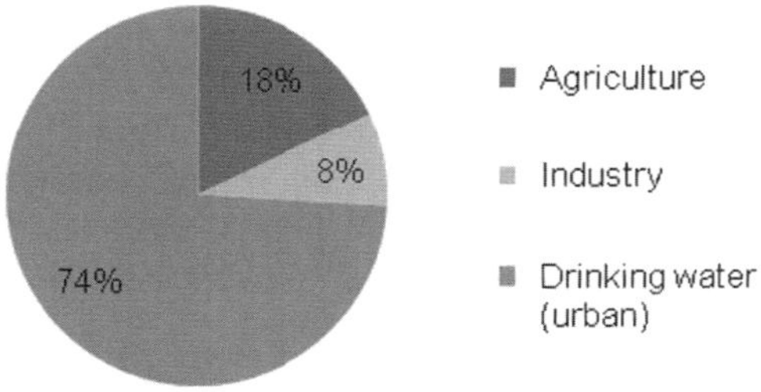

Fig. 4.4 Current share of water demand differentiated by sectors in the MRS (*Source*: own calculation based on SISS 2010; PUC 2011)

considerably from 11.4 % in the Aguas Manquehue network to 42.9 % in the SMAPA network. The average loss in all networks of the MRS was 31.2 % in 2010 (SISS 2011). Since a volume of 253.5 m. m^3 was lost in the network in that year, the four water companies were obliged to produce a total of 790 m. m^3 to provide the end user with 536.5 m. m^3 (SISS 2011).

Information on industrial water use is rare. The MRS is dominated by the manufacturing industry; in 2004, 30.24 % of all manufacturing in Chile was located in this area (INE 2005). The dominant industries are food, textiles, paper, chemicals, metal, rubber and plastic. Industrial water demand has been on the increase since 1990, reaching approximately 0.32 km^3/a (10 m^3/s) in 2006 (DGA 2007a). Since most industries have their own water sources, they are independent of drinking water companies.

4.4 Future Water Demand and Availability

4.4.1 Combining Climate Scenarios with Socio-Economic Scenarios

Similar to Chap. 3 of this book, a scenario-based approach is applied here to address climate change adaptation planning under future hydrological conditions in the light of uncertainty. As water availability heavily depends on the regional climate, the focus was on downscaling the A2 and B1/B2 scenarios referred to in the IPCC IV report (see Chap. 2). In addition to climate change, such variables as demography, economic development and types of consumption play a major role in the further development of the MRS. In this sense, two socio-economic scenarios were

Table 4.2 Specific features of the four largest water and sanitation companies in the MRS (2010) (*Source*: SISS 2011)
Note: Three companies use ground and surface water, SMAPA uses ground water only

Company	Water supply network km	Production capacity (max.) m^3/s	No. of people served	Per capita consumption in 2010 l/cap. *d	Per capita consumption in 2010 m^3/cap. *a	Consumption m. m^3/a	Water losses %	Production m. m^3/a
Aguas Andinas	11.67	30.9	5,642,630	201	73.37	414.0	31.6	605.3
SMAPA	1.5	4.3	720,903	181	66.07	47.6	42.9	83.4
Aguas Cordilleras	1.2	5.3	378,446	469	171.2	64.8	15.1	89.9
Aguas Manquehue	221.0	0.9	34,336	805	293.8	10.1	11.4	11.4
Total	235.4	41.4	6,776,315			536.5		790.0

designed for the MRS: 'Business As Usual' (BAU) and 'Collective Responsibility' (CR) (see Chap. 3). Here future water demand depends on the specific socio-economic, institutional and technical development in the MRS. Adopting the BAU and CR scenarios allows for a perspective on the state of the MRS in 2050. The time span from 2045 to 2065 is used for climate predictions (A2, B1/B2). According to these predictions, A2 can be interpreted as the 'worst case' and B1/B2 as the 'best case'—bearing in mind sustainable development for the future. Since climate change is strongly interlinked with water supply and demand, a combination of framework scenarios (Chap. 3) and climate predictions was subjected to a qualitative assessment. Assuming a smaller population, the installation of water-saving technologies in households and agriculture, and rehabilitation of the water pipe system, water demand in the CR scenario is remarkably low compared to the BAU scenario. The results of the A2 climate scenario were related to those of the BAU scenario to produce a 'worst case' model for both supply and demand. The 'best case' was formulated by combining the results of B1/B2 with the CR scenario. Using these combinations the potential water balance range can be described for the future and the extent of the challenges indicated. Estimated values for future climate and hydrological conditions (A2 and B1/B2) were taken from the work of Cortés et al. (2012, cf. Chap. 2) and used to estimate the water supply side for the time span 2045–2065. Demand estimates were made by contextualizing the socio-economic BAU and CR scenarios to the specifics of the water sector.

4.4.2 Future Water Availability

Future water availability in the MRS is influenced by the regional impacts of global climate change, notably shifts in temperature, precipitation and/or run-off rates. The Maipo river run-off tends to decrease as much as 40 % in both climate scenarios in the summer months (Chap. 2). Precipitation likewise decreases, with a projected reduction in A_{2050}[2] of minus 10–30 %, whereas in B_{2050}[3] no significant decrease is assumed (see Fig. 4.4).

Climate scenarios also show a decline in water availability. On the whole, the expected future water availability is lower in scenario A_{2050} than in B_{2050}. The comparison states 1.03 km^3 less for A_{2050} and 0.65 km^3 less for B_{2050} in normal years compared to today. Even in the B_{2050} climate scenario, which has less water reduction than A_{2050}, the lower availability of water corresponds to almost three times the storage capacity of the El Yeso reservoir.[4] As described in Sect. 4.4.1, the results of climate and BAU/CR scenarios are combined in order to identify possible future relationships between natural water supplies and water demand for the MRS.

[2] Synonym for downscaled climate scenario A2 for the time span 2045–2065.

[3] Synonym for downscaled climate scenario B1/B2 for the time span 2045–2065.

[4] Reservoirs like 'El Yeso' are crucial to the drinking water supply in the MRS, particularly in terms of dry months.

Table 4.3 Water-relevant parameters of the two scenario alternatives for the year 2050 (*Source:* Authors' own water-specific development based on framework scenarios: Barton et al. 2011a, b)

Business as usual (BAU)	Collective responsibility (CR)
The MRS in 2050 is characterized by consequences of continuing recent trends in population, economic growth, urbanization, technology and human behaviour. BAU assumes that current market-based policies remain and environmental health and ecological integrity are of less interest.	The MRS in 2050 is dominated by a strong state presence and market regulation. Environmental protection, social justice and equity are major political goals. Slower economic growth and the introduction of clean and resource efficient technologies are key determinants. CR assumes less population growth and decentralization processes that change the urban planning processes.
Urban development	
• Urban population of 8.5 million • Introduction of new technologies, but without focus on water saving • Decreasing agricultural and irrigated area • Access rate to water and sanitation 99 % • More paved areas → more flooding • Irrigation efficiency in agriculture approx. 60 % • Urban water demand is 180 l per capita and day • Tertiarization processes increase in addition to industrial sector	• Urban population of 7.9 million • Irrigation of private gardens and per capita demand is reduced by new water-saving technologies in the household and for irrigation including grey water re-use • Constant agricultural and irrigated area • Access rate to water and sanitation 99 % • Growth of green spaces → increase of living quality, better air quality • Irrigation efficiency in agriculture approx. 75 % • Urban water demand is 150 l per capita and day • Industries grow in balance with eco-systems
Institutional framework	
• Privatization with monopolistic structures • Weak government influence • Conflicts between water users caused by private water rights	• Stronger state presence and regulation by government institutions • Water saving technologies promoted • New water law is implemented
Infrastructure	
• Insufficient maintenance of water infrastructure – increase of water losses (40 %) • Multiple use of water is uncommon • Combined sewer system (waste and storm water) without retention basins – sewage often pollutes rivers	• Improvement of water infrastructure – water losses decrease (10 %) • Sewage used as resource (e.g., warmth for buildings; grey water for irrigation, treated waste water for agricultural irrigation) • New technologies at treatment plants – safe environmental disposal • New urban areas are equipped with semi-decentralized water infrastructure

Table 4.3 summarizes the key aspects of the two socio-economic scenario alternatives BAU and CR, contextualized in accordance with the overall framework for the water sector.

According to the two scenarios presented and their specific socio-economic and technological development, changes are expected in the drinking water supply and in the water demand from the agricultural and industrial sectors. Key determinants are the advancement of agricultural irrigation efficiency by implementing new technologies, the development of agricultural and irrigated areas, population growth and public policies. Data for the current water demand is combined with assumptions about changes in population size, irrigation efficiencies, water-saving efforts and rehabilitation of the pipe network in accordance with the two socio-economic scenarios (Table 4.3), allowing for quantitative assessments of future water demands.

Both scenarios show evidence of a decrease in the total water demand, notably as a result of a lower demand for water in agriculture. Figure 4.5 shows the water demand values in 2011 and for both future scenarios.

4.4.3 Future Balance of Water Availability and Demand

Both water availability and water demand will undergo change in the MRS in the future due to climate change effects and demographic, economic and technological adjustments. In line with the shifts in precipitation and stream flow (compare Chap. 2), renewable water resources in the Santiago Basin will be more modest than they are today, varying between 1.9 and 9.5 km^3 per year. The difference between the two climate scenarios with regard to average availability is comparatively small, with A_{2050} averaging 0.4 km^3 less than B_{2050}.

The total water demand in both socio-economic scenarios (BAU and CR) is in decline (see Table 4.4). Calculations are based on the assumption that efforts have been made in both scenarios to introduce new water-related technologies (according to Table 4.3), notably in agriculture. The introduction of such technologies, however, often coincides with an additional demand for a supply of electricity (e.g., drip irrigation pumps for agricultural use or systems to press waste water through membranes), which should be based on renewable energy sources. In the year 2050, the total water demand accounts for approximately 60 % of the total availability in an average year. The dry years pose the greatest challenge here: water demand exceeds availability by far (between 150 and 180 %). The key issue is how to enable the MRS to maintain a sustainable water balance.

4.5 Designing Adaptation Measures

4.5.1 The Need for Climate Change Adaptation Measures

In addressing climate issues, climate change management focuses heavily on the mitigation of anthropogenic greenhouse gas emissions. Since current emissions will impact on the severity of climate change in future years (Hulme et al. 2002), it is now widely accepted that mitigation alone will not suffice to deal with the adverse effects of climate change already occurring (Kurukulasuriya and Rosenthal 2003). To a great extent climate change is water change (IPCC 2008). The interference of climate change and the ENSO phenomenon affects drinking water security and agricultural productivity, especially in tropical and Mediterranean regions such as the MRS (Ropelowski and Halpert 1996; Kurukulasuriya and Rosenthal 2003). Consequently, the water supply in the Santiago Basin is highly mutable. Water supplies in terms of

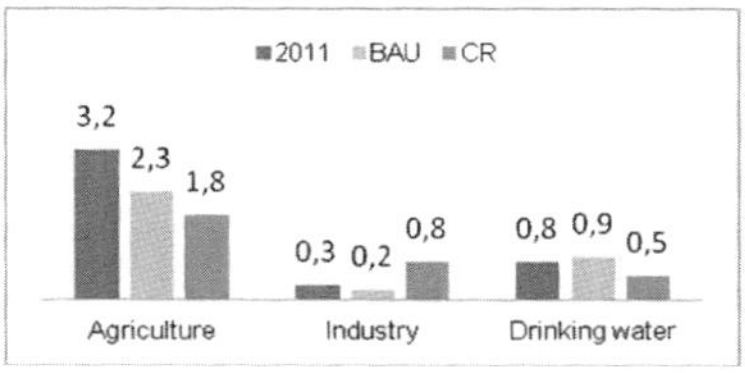

Fig. 4.5 Current annual water demand and projected annual water demand in 2050 for the MRS in km^3 (*Source*: own calculation based on framework scenarios)

Table 4.4 Relationship between future renewable water resources (according to climate scenarios) and demand (according to socio-economic scenarios—see Table **4.3**) in km^3 (*Source*: Authors' own calculation)

		2011	A $_{2050}$	B $_{2050}$
Supply	Min	2.8	1.9	2.1
	Max	10.2	9.3	9.5
	Average	**6.2**	**5.2**	**5.6**
Demand		**2011**	**BAU**	**CR**
	Total	**4.3**	**3.4**	**3.1**

security, productivity and sustainability can be assessed by the by the use of indices. Two of such indices are presented in the following.

4.5.1.1 Assessment According to the Falkenmark Index

The Falkenmark Water Stress Index is a helpful tool to estimate water availability in the Santiago Basin with respect to the number of people to be served and is broadly accepted in the water community (Falkenmark 1989). The index covers basic water demands: drinking water, agricultural, and industrial production. Four water availability categories are defined:

- Sufficient water: >1,700 m^3/cap.*a
- Water stress: <1,700 m^3/cap.*a
- Water scarcity: <1,000 m^3/cap.*a
- Severe water scarcity < 500 m^3/cap.*a

The current overview of available water in the Santiago Basin (407–1,483 m^3/cap*a, cf. Table 4.1) indicates that there is already a scarcity of water, with 'severe water scarcity' in dry years and 'water stress' in wet years.

4.5.1.2 Assessment According to the Water Exploitation Index

Whereas the Falkenmark Index assesses the water situation in the context of the number of people to be served, the Water Exploitation Index (WEI) calculates the ratio between water extraction and renewable water resources, marking the relationship between the natural supply potential and societal needs (Marcuello and Lallana 2010). A WEI >1 indicates that a greater volume of water is extracted than is renewed. In the long run a balance between extraction and renewal will only be possible if sufficient water storage capacities are available. In view of the figures for water demand and annual renewable water resources (see Table 4.4), the WEI currently assesses the Santiago Basin as having medium water stress (0.42) in wet years, high water stress (0.69) in normal years, and very high water stress (1.54) in dry years—according to (Döll 2008) (Fig. 4.6).

Groundwater in the Santiago Basin is an ideal natural facility for water storage: the underground location of this water resource affords greater protection against pollution and warming up compared to surface waters. Due to the slow speed of groundwater (as distinct from rivers), its residence time in the region is higher, thus enabling water storage during periods with little precipitation. The management of this pivotal water reservoir in the Santiago Basin, however, cannot be assessed as sustainable: the constant lowering of the groundwater table (from 12 to 26 m below the surface) in the period from 1969 to 2001 (DGA 2007b) clearly indicates that water extraction exceeds the rate of natural groundwater recharge (Fig. 4.7).

The two indices show that the Santiago Basin is already suffering from water stress and water scarcity in both normal and dry years. The water demands of the population can only be met by neglecting the water demands of nature or of ecological services, and exploiting natural water storage capacities beyond their ability to recharge.

These findings are evidence of the current urgency for demand-related measures to achieve a balance between natural water availability, on the one hand, and the demands of the environment and human society, on the other. Taking into account the results of the water balance assessment, as well as the impacts of climate change, and demographic and economic developments (see Sect. 4.4.3), the water situation in the Santiago Basin will deteriorate in the future if adaptation measures will be waived.

4.5.2 Water-Related Measures to Adapt to Climate Change

In order to respond to the already traceable problems identified in the water sector of the MRS, which will more than likely be aggravated in the future, adaptation measures need to be designed and implemented. The measures presented in this chapter are a combination of stakeholder knowledge in the scenarios, scientific expertise and existing policy guidelines. They were identified, refined and prioritized from a set of possible measures drawn up in a participatory process (see Chap. 9). Generally, adaptation to change in the water sector can be subsumed under institutional measures and measures related to supply and demand. Four of the wide range of measures proposed were studied in detail by Chilean experts, who conducted scoping studies to assess their feasibility in the MRS. Selection of the four measures was primarily based on their effectiveness and/or potential to save water:

1. Water demand in the MRS is highest in the agricultural sector. Upgrading the effectiveness of irrigated agricultural systems has enormous water-saving potential.
2. Reducing the water demand in private households has the positive side effect of increasing public environmental awareness.

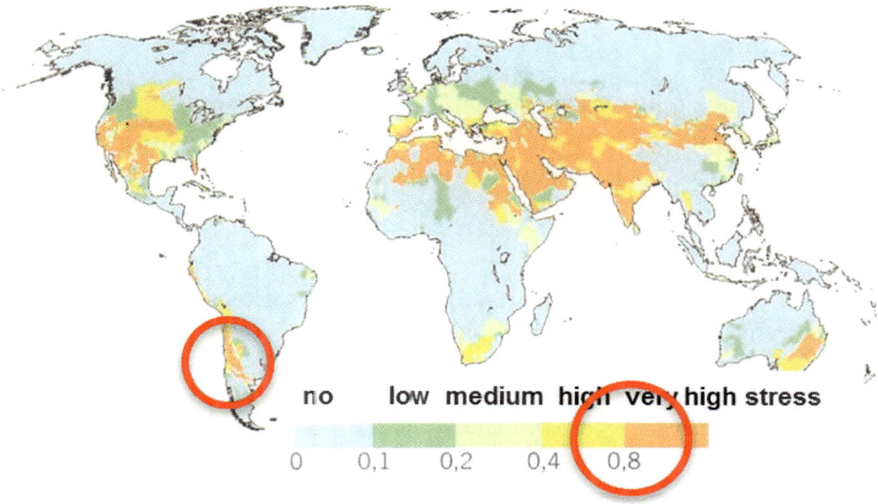

Fig. 4.6 Water stress in river catchments in the year 2000 according to the Water Exploitation Index (WEI) (*Source*: Döll 2008)
Note: A catchment value of more than 0.4 indicates high water stress and WEI>0.8 very high water stress

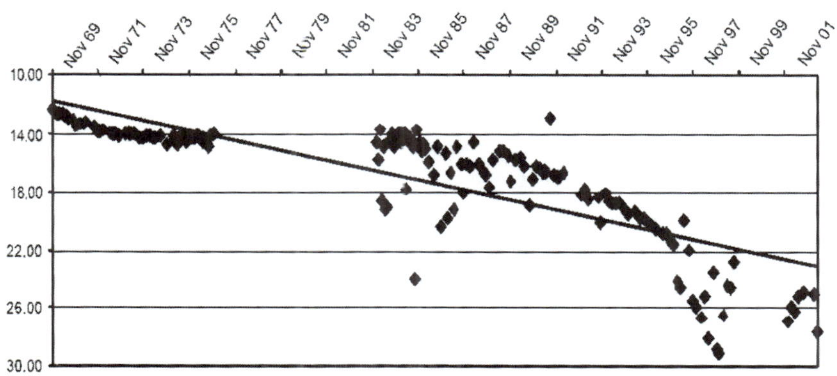

Fig. 4.7 Groundwater table from 1969 to 2001 within the Sector Santiago Central at Estación Consejo Nacional de Menores (*Source*: DGA 2007b)

3. Re-using grey water systems in new built-up areas also contributes to public awareness.
4. Implementation of a new policy structure for the water sector is recommended in order to embed measures in the changing institutional background.
   The four measures are described in more detail below.

4.5.2.1 Reduction of the Total Agricultural Water Demand by Introducing Water-Efficient Irrigation Technologies

According to the latest agricultural census, irrigation efficiency[5] in the MRS is in the order of 36 %, which is quite low compared to international standards (INE 2007; PUC 2011). The agricultural sector in Chile could therefore make great strides to save water. The proposal for this measure is based on the assumption that the irrigated agricultural area will remain constant in the future and that irrigation efficiency can be improved, ultimately leading to compensation for reduced water availability resulting from the adverse effects of climate change. Hence the utilization of efficient irrigation technologies such as pressurized water application techniques (sprinkler and micro irrigation) (Kulkarni 2011) should be accompanied by measures to ensure that the water saved is not transferred to other agricultural areas or other sectors (e.g., mining). This could be achieved by introducing a decree to prohibit the allocation of new irrigated agricultural areas in the MRS. It is furthermore proposed that subsidies for new irrigation technologies should be dedicated to areas currently equipped with low-efficient technologies. Financial support to install specific systems with at least 75 % efficiency, such as sprinkler or drip irrigation systems,[6] should be restricted to recipients who voluntarily return to the state the volume of water they save (in terms of water rights). A bonus system should be established to motivate farmers to yield their unused water rights to the state (as a civil law contract). Expected co-benefits are reduced water costs for farmers due to lower water consumption, less exploitation of groundwater resources and a water flow rate that maintains healthy aquatic ecosystems (ecological flows). Identification of the main agricultural areas with low-efficient irrigation technologies—a priority in terms of modernization—should precede the measure.

An irrigation subsidy programme was established in Chile by the National Irrigation Commission (CNR) (CNR 2009). It seems unnecessary to implement a new funding scheme. Instead, efforts should be made to improve existing agricultural policies. The policies in place focus mainly on increasing agricultural productivity and contain no incentives to reduce water demand, a prerequisite for long-term sustainable production. Neither do existing policies include incentives to seek alternative water resources or uses (e.g., water re-use) or to strengthen the capacity of user associations. As a rule these policies are highly inflexible. It is therefore recommended that future agricultural subsidy policies guarantee the stability of the total agricultural area to be irrigated.

Actors from a number of government institutions, such as CNR, the Ministry of Environment (MMA), the National Office for Regional Development (SUBDERE) and the Institute of Agricultural Development in Chile (INDAP), should be

[5] 'Irrigation efficiency' is defined as the relationship between the amount of water plants require for optimum growth and the amount of water brought to the field by irrigation (INE 2007).

[6] The INE (2007) defines efficiencies as follows: 75 % for sprinkler irrigation, 85 % for drip irrigation and 90 % for micro-irrigation.

involved in the implementation of this measure. A barrier to its successful implementation could be the expected lack of cooperation by farmers. It is assumed that the latter fear financial cuts following a potential drop in agricultural production. Hence gaining their confidence and participation, and ensuring their financial needs is of the utmost importance.

4.5.2.2 Introduction of Water-Efficient Fixtures in Private Houses

The current per capita water demand in the MRS of approximately 220 l per day is comparatively high (SISS 2011). The aim should be to reduce the future water demand to about 150 l/cap-day, i.e., 100 l/cap-day for household purposes and 50 l/cap-day for the irrigation of private green spaces under Mediterranean climate conditions. In 2004, for example, the per capita water demand in 88 cities and villages of the south-western German state of Baden-Württemberg was less than 100 l/cap-day (Bühringer 2006). This clearly indicates that even today 100 l/cap-day is an achievable level for existing buildings. According to the Bayerische Landesanstalt für Weinbau und Gartenbau (Bavarian Regional Office for Viniculture and Horticulture) (2007), one square metre of high quality lawn requires 3.3–4 l of water under German summer conditions. Assuming that the latter are akin to weather conditions prevailing throughout the year in the MRS, 50 l of water should suffice to irrigate a lawn of 12–15 m^2 in size. In other words, with the technologies already in operation in Chile, 150 l/cap-day should cover household requirements, including a green area around the house. To presume that flowering plants need less water than lawns and that the reported consumption of less than 100 l/cap-day in German cities includes a certain amount of irrigation water illustrates the conservative character of this assessment. The introduction of water-efficient tap fittings in bathrooms or kitchens, and of efficient toilet-flushing systems in existing buildings reduces water consumption at a comparatively low cost. Implementation of this measure would consequently reduce water and sewage bills. Lower water consumption (e.g., while showering) likewise leads to lower energy-related costs. Additionally, more efficient water use combined with water-saving technologies could serve to avoid expensive investment costs otherwise needed to adjust water supplies and waste water treatment facilities to a growing population. Lowering human water consumption with water-saving measures is a prerequisite to assuring ecological flows in natural water bodies and thus to guaranteeing water-based environmental services—particularly in water scarce regions like the MRS.

Experience in Germany shows that the per capita water demand of hotels is higher than that of private households (Lehn et al. 1996). The installation of water-saving fixtures in hotels could reduce the per capita water demand even more efficiently than in private households. Since the high per capita water demand in hotels results from intensive showering, saving water here will lead to energy saving when the demand for warm water declines.

The overall aim of the measure is to:

- Reduce approx. 30 % of urban water demand by introducing water-efficient installations up to 2050.

- Draw up strategies for existing and new houses: (1) Gradual exchange of installed sanitary fixtures in existing houses accompanied by awareness-raising campaigns and economic incentives when the pay-back period exceeds 1 year. (2) Setting obligatory water efficiency standards for sanitary systems.
- Create an efficiency label for a range of water efficiency products (percentage of water saving) and limit the maximum flush volume for each of these fixtures (with readjustment every five years) as the basis for incentives.
- Establish an adequate subsidy system: depending on the price of water-saving fixtures and the attendant pay-back periods, subsidies could be made available for the refitting of installations in existing buildings.

The expected co-benefits are energy saving due to less (warm) water, reduced water and energy costs, and growing public awareness of the need and opportunity to save water. Since the amount of soap or shower gel is expected to remain constant with water-saving fixtures and the composition of excreta is unlikely to change with water-saving toilets, the relation between water and its ingredients will result in higher concentrations of ingredients in waste water. Thus waste water treatment plants are expected to achieve greater efficiency as a result of less diluted sewage.

4.5.2.3 Treatment and Re-use of Grey Water in New Built-Up Areas

Approximately 19 % of the total MRS superficies is vegetal material in the form of lawns, plants, bushes and trees (Moya 2009). The water demand in this area is approx. 250,000 m^3/day. According to a survey[7] carried out by the Observatorio de ciudades OCUC, 60 % of consulted households were prepared to improve their irrigation systems with more efficient, water-saving technologies, while around 46 % would agree to irrigate with treated grey water[8] (OCUC 2010). Grey water recycling is an appropriate measure to adapt water management to the impacts of climate change, and its gradual integration at all levels of legislation, planning, construction and management of urban green areas is the overall aim of the measure. Successful grey water management includes both technical methods, and institutional and legal aspects (e.g., user participation in planning, running and maintaining the systems). The overall objective of these measures is:

- in the short run to raise awareness among architects, planners and investors of the interrelation between the drinking water supply and the re-use of grey water and its potential for irrigation of green spaces when they plan green areas, building dimensions and technical installations.
- to establish a water quality norm for recycled grey water and a technical guide for the certification of domestic grey water systems for irrigation needs.

[7] Three hundred households in 15 different communities were consulted between 4 and 27 June 2009. The survey was conducted with people over 18 years of age from different socio-economic backgrounds.

[8] Generally, grey water includes all household waste water with the exception of toilet and kitchen effluents.

- in the long run to substitute drinking water with treated grey water for irrigation of urban green spaces.

A key issue in the planning process is synchronization of the amount of recycled grey water with the availability of irrigated areas (size, water demand of plants) to adopt grey water for irrigation.

Expected co-benefits for the future are the saving of drinking water and reduced water bills, optimization of waste water treatment plants due to less dilution of waste water, and the enhanced environmental image of Santiago de Chile. In a first step, this measure should be realized with pilot projects in newly built housing areas. Capacity development and education campaigns to raise awareness of the topic (notably, for example, among architects, investors, planners) are of the utmost importance for the success of this measure. Further, implementation will require determining an institution to be responsible for grey water recycling (ability to lead and coordinate existing regional policy and the institutions involved) and a suitable financing facility to support the implementation of grey water recycling systems. It is impossible to specify the installation costs of a grey water recycling system, since the complexity of treating grey water depends on its quality and on the specific system and its design.

If the relation between built-up areas and green spaces is aptly designed, storm water can be seeped decentrally onto the latter. The grey water infrastructure can also be used to transport storm water from the built-up area to green spaces. Since storm water requires pre-treatment, its inclusion in the grey water system is feasible. In this case, storm water collectors would no longer be necessary and the attendant costs saved.

Obstacles can be expected: the current lack of awareness and knowledge of these issues among the authorities, investors, planners and inhabitants needs to be overcome. The introduction of economic incentives depends on the public budget and the economic stability of public households.

4.5.2.4 Establishment of Integrated Governance Structures for the Maipo-Mapocho Watershed

The pressure on water resources caused by climate conditions, population growth, increasing water demand and water pollution highlights the hydrological, social, economic and ecological interdependencies in the Maipo-Mapocho watershed. This calls for an integrated management system involving the basin stakeholders concerned (private and public). The aim of the measure is to set up appropriate administrative structures. As a rule, the public authorities remain the final supervisory level. Private stakeholders are more 'involved in decision-making processes' than in 'making decisions'. Hence the establishment of two entities is recommended:

1. Regional Water Council: a political institution that seeks to bring the different public and private actors together. It could play an active part in shaping and accompanying the dialogue and participation process for the adaptation strategy. This would ensure a consistent conceptual approach by the regional government.

2. River Basin Entity: an executive board responsible for coordinating the relationship between the various services and hydrological planning in the Maipo watershed. The integrated regulation and control of all water-based needs in the catchment according to sustainability principles, e.g., of water extraction in relation to available supplies, of water quality for different purposes, and of the promotion of sustainable water-use practices, could be some of the concrete activities of this body. Apart from public participation in decision-making processes, expected co-benefits are the reinforcement of regional administrative structures and enhanced participation by the stakeholders concerned.

4.6 Conclusions

From a historical point of view, the founders of the city of Santiago made a fitting choice of place to settle as far as water supply and sewage disposal is concerned. The complementary character of rainfall in the winter months and the inflow in spring, summer and autumn of melt waters from vast natural water storage facilities (glaciers and snowfields) in the high Andes guaranteed a water supply throughout the year. As a result of the steep slopes of the western valleys in the Andes, the velocity of river waters was high, allowing sewage to be discharged rapidly downstream from the city to the Pacific Ocean.

Over time the inhabitants altered these favourable framework conditions, leading ultimately to the more risky situation that prevails today. Due to the growth in population and increased economic activities all of the available water resources have been allocated to a specific use. Water storage capacities have been augmented by the construction of dams—notably the *El Yeso* dam—to overcome dry annual periods and dry (La Niña) years in particular. Despite these efforts, the Maipo riverbed dries up regularly in the summer months after it leaves the city, and the groundwater table has been subsiding for decades. From an integrative perspective and taking into account ecological services and/or a balanced relationship between upstream and downstream riparians, this river basin management is clearly unsustainable. Measures to adapt water-use patterns to the water supply regime are thus crucial even today.

The interrelationship between climate change effects and the ENSO phenomenon—with increasing intra-annual water supply variations—will in the future lead to a widening of the gap between water availability and water demand. Hence, there is an urgent need for adaptation measures if depletion of water resources in the MRS is to be avoided. The vast potential to reduce water demand without lowering the quality of life has been demonstrated. Adaptation measures should be taken into consideration on the demand and the supply side in both urban and rural areas, and involve residents, farmers, industries and water supply companies. Technical, institutional and behavioural measures are proposed in order to engineer integrative and adaptive water management.

As far as the demand side is concerned, the greatest water-saving potential exists in the agricultural sector, since irrigation efficiency is currently at a very low level.

In line with the scenarios and the specific climate setting in the MRS, raising efficiency to international standards would mean saving between 0.9 and 1.4 km^3 of water or between 900 and 1,400 m. m^3. In contrast, the actual total drinking water demand in the urban area of Santiago accounts for 700–800 m. m^3 (see Fig. 4.5). Upgrading efficiency alone, however, will not suffice. Technical measures in the field of irrigation must also be flanked by an improved regulation scheme, which in turn guarantees that the volume of water saved is not wasted in other sectors or regions. In addition, the urban sector has considerable potential to save water on the demand side. Yet another adaptation measure is the improvement of the drinking water network, which was not discussed in depth during the participatory process (see Chap. 9).

According to the calculations described in Sect. 4.4, drinking water losses of approx. 175 mil. m^3/a could be avoided if the leakage rate were reduced to 10 %. This is an ambitious but achievable target, as seen in the case of Aguas Manquehue (see Table 4.2), which reports a leakage rate of 11 %. In the urban area, an annual saving potential of approx. 350 m. m^3 seems realistic and can be achieved without the introduction of new technology. What is merely required is the installation of water-saving fixtures and the rehabilitation of the pipe network. The application of new methods and technologies, such as grey water recycling, could lead to an estimated annual saving of a further 100–120 m. m^3 of drinking water.

The climate scenarios indicate that water supplies could decrease in the year 2050 by about 1 km^3 or 1,000 bn. m^3 (see Table 4.4). Mobilizing the above-mentioned potential could mean saving more than 1,500 m. m^3 of water in the Santiago Basin, leaving a slightly higher volume of water for ecological services to nature. This calls for a new integrated governance structure for the Maipo watershed, which would control and regulate all water-based needs in the catchment in a manner that is both comprehensive and in compliance with the principles of sustainability.

The proposed technical and institutional adaptive measures are at risk of failing if public awareness of water issues remains at its current level. The high per capita consumption in households (especially in high-income areas), exorbitant losses in the pipe network and low irrigation efficiency in the agricultural sector and urban domestic garden watering clearly indicates that large sections of society are unaware of the current dramatic imbalance between natural water supplies and anthropogenic water demands. To counteract this deficit a structured and long-term awareness campaign is recommended.

Overcoming the gap between water supply and water demand in the Metropolitan Region of Santiago poses a huge challenge today and even more so in the future. Santiago's natural features and its ability to cope, however, give reason to hope that this challenge will be addressed.

References

Barton, J., Kopfmüller, J., & Stelzer, V. (2011a). *Project 'ClimateAdaptationSantiago' (CAS): Quantitative variables: Tendencies and scenario estimations 2010–2050*. Working Paper.

Barton, J., Kopfmüller, J., & Stelzer, V. (2011b). *Project 'ClimateAdaptationSantiago' (CAS): Scenario storylines for the RMS*. Working Paper.

Bartosch, A. (2007). *Die Wasserversorgung in einer Metropolregion in Lateinamerika. Das Beispiel Santiago de Chile*. Thesis. Universidad Friedrich-Schiller-Universität Jena.

Bayerische Landesanstalt für Weinbau und Gartenbau. (2007, August). *Rasen und Wiese im Hausgarten*. Veitshöchheim, leaflet.

Biblioteca del Congreso Nacional de Chile Sistema integrado de Información territorial (SIIT). (2013). *Division regional: Polígonos de las regiones de Chile*. http://siit2.bcn.cl/mapas_vectoriales/index_html/. Accessed 22 Feb 2013.

Bühringer, H. (2006). Trinkwasserversorgung in Baden-Württemberg. *Statistisches Monatsheft Baden-Württemberg, 5*, 28–31.

Comisión Nacional de Riego (CNR). (2007). *Diagnóstico de Caudales, Disponibles en Cuencas No Controladas de Recuperación*. Cuencas Aconcagua y Maipo. Santiago de Chile.

Comisión Nacional de Riego (CNR). (2009). *Ley Nº 18.450 de Fomento a la Inversión Privada en Obras de Riego y Drenaje*. Santiago de Chile.

Cortés, G., Schaller, S., Rojas, M., Garcia, L., Descalzi, A., Vargas, L., McPhee, J. (2012). Assessment of the current climate and expected climate changes in the Metropolitan Region of Santiago de Chile. UFZ-Report 03/2012, Helmholtz Center for Environmental Research UFZ, Leipzig.

Dirección General de Aguas (DGA). (2007a). *Estimaciones de Demanda de Agua y Proyecciones Futuras. Zona II. Regiones V a XII y Región Metropolitana*. DGA Publicación S.I.T. Nº123. Santiago de Chile.

Direccíon General de Aguas (DGA). (2007b). *Evaluación de la explotación maxima sustentable del acuífero Santiago sur*. Santiago de Chile.

Dirección Meteorológica de Chile (DMC). (2012). *Anuarios climatológicos 1960-2011*. Santiago de Chile.

Döll, P. (2008). Wasser weltweit – Wie groß sind die globalen Süßwasserressourcen, und wie nutzt sie der Mensch? *Forschung Frankfurt, 3*, 54–59.

Falkenmark, M. (1989). The massive water scarcity threatening Africa – Why isn't It being addressed? *Ambio, 18*(2), 112–118.

Hulme, M., Jenkins, G. J., Lu, X., Turnpenny, J. R., Mitchell T. D., Jones, R. G., et al. (2002). *Climate change scenarios for the United Kingdom: The UKCIP02 scientific report*. Tyndall Centre for Climate Change Research, School of Environmental Sciences, University of East Anglia, Norwich, UK.

Instituto Nacional de Estadísticas. (2005). *Región Metropolitana. Perfil de la dinámica económica regional*. Santiago de Chile.

Instituto Nacional de Estadísticas (INE). (2007). *VI Censo Agropecuario Nacional, 2007*. Santiago de Chile.

Instituto Nacional de Estadísticas. (2008). *División político-administrativa y censal, 2007*. Santiago de Chile.

Intergovernmental panel on climate change. (2008). *Technical paper on climate change and water*. http://www.ipcc.ch/pdf/technical-papers/climate-change-wateren.pdf. Accessed 30 July 2012.

Kopfmüller, J., Lehn, H., Nuissl, H., Krellenberg, K., & Heinrichs, D. (2009). Sustainable development of megacities: An integrative research approach for the case of Santiago Metropolitan Region. *Die Erde, 140*(4), 417–448.

Kulkarni, S. (2011). Innovative technologies for water saving in irrigated agriculture. *International Journal of Water Resources and Arid Environments, 1*(3), 226–231.

Kurukulasuriya, P., & Rosenthal, S. (2003). *Climate change and agriculture: A review of impacts and adaptations world bank climate change series paper no. 91*. Paper prepared and published for the Rural Development Group and Environment Department of the World Bank.

Lehn, H., Steiner, M., & Mohr, H. (1996). *Wasser – die elementare Ressource*. Heidelberg: Leitlinien einer nachhaltigen Nutzung.

Lehn, H., McPhee, J., Vogdt, J., Schleenstein, G., Simon, L.-M., Strauch, G., et al. (2012). Risks and opportunities for sustainable management of water resources and services in Santiago de Chile. In D. Heinrichs, K. Krellenberg, B. Hansjürgens, & F. Martínez (Eds.), *Risk Habitat Megacity* (pp. 251–278). Heidelberg: Springer.

Marcuello, C., & Lallana, C. (2010). *Indicator fact sheet (WQ01c) water exploitation index*. http://www.eea.europa.eu/data-and-maps/indicators/water-exploitation-index. Accessed 31 July 2012.

Meza, F. J. (2005). Variability of reference evapotranspiration and water demands. Association to ENSO in the Maipo river basin, Chile. *Global and Planetary Change, 47*, 212–220.

Moya, L. (2009). *Contribución de los Jardines Domésticos Urbanos a la Cobertura Vegetacional de Santiago de Chile*. Tesis presentada al instituto de Estudios urbanos y territoriales de la Pontificia Universidad de Catolica de Chile, Santiago de Chile.

Obervatorio de ciudades (OCUC). (2010). *Formulación sello de efficiencia hídrica en el paisaje*. Santiago de Chile.

Pontificia Universidad Católica de Chile, Centro de Cambio GLOBAL (PUC). (2011). *Analysis of agricultural water demands in the Maipo Basin*. Technical report. Santiago de Chile.

Ropelowski, C. S., & Halpert, M. S. (1996). Quantifying Southern oscillation-precipitation relationships. *Journal of Climate, 9*(5), 1043–1059.

Superintendencia de los servicios Sanitarios (SISS). (2010). *Informe de gestión 2009*. Santiago de Chile.

Superintendencia de los servicios Sanitarios (SISS). (2011). *Informe de gestión 2010*. Santiago de Chile.

Climate Change Impacts on the Energy Sector and Adaptation Options

5

Volker Stelzer and Adriana Quintero

Abstract

This chapter demonstrates the future impacts of climate change on the energy sector in the Metropolitan Region of Santiago de Chile (MRS) and develops measures to reduce problems ahead. The investigation is based firstly on an overview of recent energy supply and demand in the MRS. Existing data taken from the literature and interviews indicates that only a small amount of the energy consumed in the MRS is produced on its territory. Secondly, this chapter makes use of a description of the framework scenarios outlined in Chap. 3, where two quantitative scenarios for the supply and demand of energy were created for the MRS. Both scenarios see an increase in the consumption and supply of energy. Thirdly, a calculation was made for the future influence of climate change on the energy supply and demand. Using a methodology of fixed correlations between rising temperatures and energy consumption, it indicates that climate change will have a relatively low impact on energy supply and demand. So dependence of the MRS on energy imports is expected to rise even further in the future. Four measures were designed for implementation in the MRS as a response to future problems.

Keywords

Energy infrastructure • Urban energy demand • Urban energy supply • Energy diversification • Energy efficiency • Santiago de Chile

V. Stelzer (✉) • A. Quintero
KIT Karlsruhe Institute of Technology, ITAS-Institute for Technology Assessment and Systems Analysis, Campus North, P.O.B. 3640, 76021 Karlsruhe, Germany
e-mail: volker.stelzer@kit.edu

K. Krellenberg and B. Hansjürgens (eds.), *Climate Adaptation Santiago*,
DOI 10.1007/978-3-642-39103-3_5,

5.1 Energy and Climate Change

The energy sector is a topic frequently mentioned in association with climate change. This is primarily because global energy supplies are responsible for approximately 68 % of CO_2 emissions (IEA – International Energy Agency 2011). Most studies and research focus on strategies to reduce greenhouse gas emissions (mitigation), e.g., by restricting future total emissions in individual countries and regions to a predetermined level (Mirasgedis et al. 2007). The potential impact of climate change on supply and demand in this sector, however, and the methods of dealing with this challenge have rarely been addressed in research and have failed to take centre stage in socio-political discussions.

The demand for energy in the Metropolitan Region of Santiago de Chile (MRS) has been on the increase for decades due to rising population and gross domestic product (GDP) growth rates, a trend that will continue according to the projected demographic, political and economic development of the MRS (Chap. 3). Meeting this demand—particularly in the area of electricity and heating—calls for expansion of the energy sector within the region and beyond. The aim of this chapter is to analyse the energy sector in the MRS for the period up to 2050 with explorative scenarios, to examine the impact of climate change on supply and demand, and subsequently to draw up suitable adaptation measures. These measures are an integral part of the overall Regional Climate Change Adaptation Plan for the MRS described in Chap. 9.

The impact of climate change on energy demand and energy supply is calculated with data on climate change in the MRS (changes in temperature, precipitation patterns and amounts, stream flows of the relevant rivers) (cf. Cortés et al. 2012 and Chap. 2). Since a large share of the electricity generation comes from hydroelectric and thermal power stations, the focus lies on the impact of climate change on these two technologies.

5.2 Present State of Energy Demand and Supply

Future scenarios and the above-mentioned measures are based on the current status of the Chilean energy market as well as on a review of the literature, data collection, calculations, expert interviews and a series of workshops.

Information required for documentation of the current energy situation in the MRS was garnered from a number of sources. These included government agencies such as the National Energy Commission (CNE), the Ministry of Energy (ME), the Ministry of the Environment, the National Institute of Statistics (INE), the Chilean Superintendence of Electricity and Fuels (SEC), and private and public companies (Chilean Electricity Company (Chilectra), National Petroleum Company of Chile (ENAP), Chilean Petroleum Company (COPEC), Chilean Distributor for Natural Gas for the Metropolitan Region (METROGAS), Power Transmission Company (Transelec)). The University of Chile (UC) and the Catholic University of Chile (PUC) likewise supplied data. Interviews conducted with stakeholders complemented the data.

Data on energy consumption is taken from such sources as SEC and INE. Consumption in the individual sectors (residential, commercial, industrial, transport) was calculated for electricity and fuels (gas and liquid). These calculations were based on several assumptions, since some of the original data, notably on liquefied petroleum gas (LPG) and liquid fuels, was not specifically allocated to the above-mentioned sectors.

LPG data was provided in two categories: LPG *sold in cylinders* (packaged) and LPG *distributed to storage tanks* (bulk). Personal interviews were held with ENAP and METROGAS staff to determine a formula that would convert the data into sectorial data. This led to the following calculation: the residential sector uses up all of the LPG sold in cylinders and ten per cent of the LPG distributed to storage tanks; the industrial sector consumes 80 % of the LPG distributed to storage tanks; the remaining ten per cent is absorbed by the commercial sector.

Data on liquid fuels was originally assigned to the following categories: users (sales to industry, commercial and residential use), transport (sales to land transport companies), ranchos (sales to aircraft operators), retail channels (sales to local service stations and retail in general) and internal consumption (distribution company vehicles). According to the National Socio-Economic Survey of the year 2006 (Ministerio de Planificación 2006, Gama Ingenieros 2009) and personal interviews with ENAP, however, only data in the "user" category is required for the calculation of consumption related to the residential, commercial, industrial and transport sectors. All of the domestic kerosene is consumed by the residential sector. The commercial sector absorbs 20 % of fuel oils and five per cent of gasoline and diesel. Industry consumes 80 % of fuel oils and ten per cent of gasoline and diesel. Finally, the transport sector swallows 85 % of gasoline and diesel and 100 % of other petroleum derivates.

No data adjustment was required to calculate the consumption of electricity, as the original data had been allocated to the four sectors (residential, commercial, industrial and transport). Statistics for mining and agriculture, which were provided separately, were added to the industrial sector.

A wealth of general information is available on the energy supply side, including data from large power stations. Information on the electricity supply is based on data from the Central Interconnected Electric System—Load Economic Dispatch Center (CDEC-SIC), the operator of the regional electricity supply system.

Section 5.2.1 gives an overview of the energy supply and demand in the residential, commercial, industrial and transport sectors of the MRS for the year 2010.

5.2.1 Energy Situation in Chile

The energy supply of the MRS is highly dependent on national and international resources. Chile's energy consumption has risen continuously in recent decades, stimulated by demographic and economic growth. This trend levelled off temporarily in 2009 (Fig. 5.1) as a result of the global financial crisis, which led to a decrease in the GDP growth rate. Following the crisis, energy consumption once again began to rise. The energy sector in Chile is privatized to a major extent

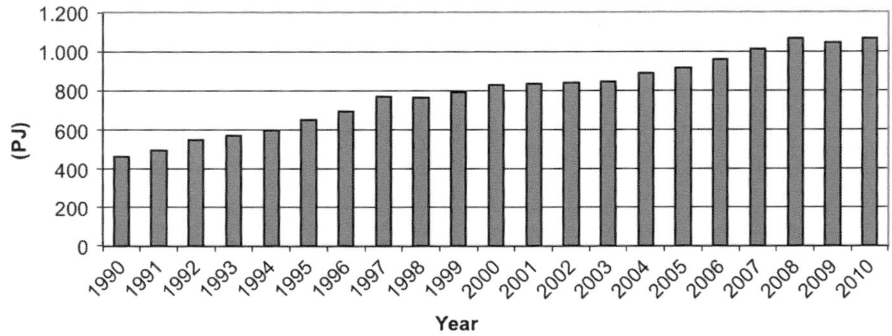

Fig. 5.1 Final energy consumption in Chile (*Source*: Authors' own presentation based on ME 2010)

and highly concentrated. The three main electricity customers hold a total market share of 98.2 % and the three principal electricity providers, 98.5 % (Simon et al. 2012). Since Chile has practically no fossil energy resources, it relies heavily on imports. In the aftermath of the energy crisis in 2007 and 2008 caused by a 90 % reduction in natural gas supplies from Argentina, Chile built two LPG terminals to minimize the risk of interruptions in the supply of energy (Gas Natural Quintero 2012).

In 2010, the electricity production in Chile had an installed capacity of 16.716 megawatt (MW) (CER 2012) and was based primarily on thermal electricity generation (63.0 %) and hydropower (26.4 %), with an added 0.6 % wind power (CDEC-SIC 2012). Recent years saw the launch of programmes to exploit the vast potential of renewable resources in Chile, e.g., wind, solar and geothermal energy. In 2011, 116 MW generating capacity was added to this, representing an increase of ten per cent. The production of 242 MW is under construction: a total of more than 1,000 MW was approved in 2011 following an environmental impact assessment, translating to an investment in excess of $2 600 m (CER 2012). Finalization of these projects will increase the production capacity for electricity in the order of seven per cent. Electricity distribution is divided into four grids, the largest of which is the CDED-SIC grid (approx. seven per cent of production) that feeds the MRS with electricity (ME 2010).

5.2.2 Current Energy Demand in the MRS

The MRS is home to 40 % of the Chilean population, which coincides with 43.5 % of the country's gross domestic product (GDP) (BCC 2011) and in 2010 accounted for 22 % of its final energy consumption (ME 2010). Given the percentage of the population and the economy that these figures represent, energy consumption in the MRS is relatively low. This is not surprising since most energy-intensive mining and paper industries are located outside the MRS in the northern and southern parts of the country.

Final energy consumption in the MRS by economic sector has shown a similar pattern in recent years. In 2010, the transport sector consumed 47 % of final energy, followed by the residential and industrial sectors with 21 and 20 %, respectively. The remaining twelve per cent was absorbed by the commercial and services sector (Fig. 5.2).

Final energy consumption in the MRS in 2010 breaks down to 51 % of liquid petroleum products, followed by 26 % of electricity and 22 % of gas (twelve per cent liquefied petroleum gas; ten per cent natural gas). Just one per cent of final energy consumption is wood based (authors' own calculation based on INE 2010; SEC 2006–2010). As climate change has in general relatively little influence on the consumption of fuels in the transport sector (Mansur et al. (2008), the analysis chosen for this chapter focuses on the demand for electricity and heat in the residential, commercial and industrial sectors.

Total electricity consumption in 2010 was 61 petajoule (PJ), with the industrial and commercial sectors responsible for approximately 36 % each, the residential sector for 27 % and transport (metro) for approximately one per cent (authors' own calculation based on INE 2010).

Per capita electricity consumption in the residential sector of the MRS varies considerably between municipalities. While the highest-income municipalities (e.g., *Vitacura*) have an annual consumption rate of approximately 1,200 kilowatt hour (kWh) of electricity per person, lower-income municipalities (e.g., *Alhué* and *El Monte*) use less than 360 kWh per person and year (Ministerio de Planificación 2006). Electricity consumption rates in the residential sector of the MRS average at around 700 kWh per person and year (Ministerio de Planificación 2006).

5.2.3 Current Energy Supply in the MRS

Electricity and heat are generated in the MRS with the exception of liquid fuels. Four-fifths (or 80 %) of the electricity to be consumed is imported through the Central Interconnected System (SIC) grid. The remaining fifth produced in the MRS is based in equal measure on local water resources and thermal power plants (ME 2010).

5.2.3.1 Generation of Hydroelectricity

All hydroelectric power stations located in the MRS are run-of-river power stations and exploit the potential energy of water from the Maipo River and its tributaries. These stations represented an installed capacity of 334 MW in 2010 (CDEC-SIC 2011), i.e., 40 % of the installed electricity capacity of the MRS. Most of this comes from large stations such as Alfalfal or Queltehues with a capacity of 321.3 MW or 96 % (Fig. 5.3). The remaining four per cent is provided by small stations with a capacity of less than ten MW.

Figure 5.3 shows that the annual energy production of each plant is not constant, but depends on the degree of utilization (hours per annum and the amount of water available for electricity production). The water volume available for electricity

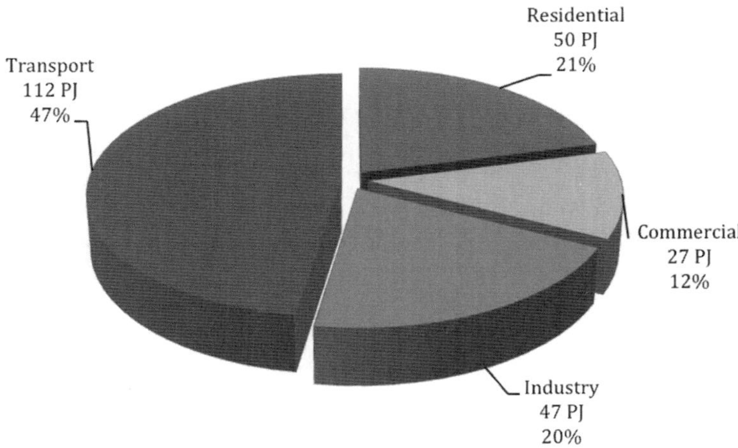

Fig. 5.2 Final energy consumption in the MRS by sector (2010) (*Source*: Authors' own presentation based on INE 2010 and SEC 2006–2010)

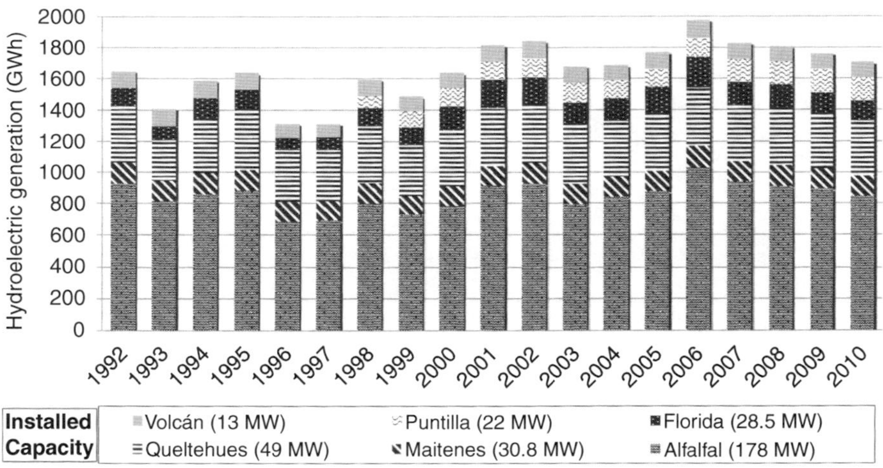

Fig. 5.3 Power in the MRS generated by hydropower plants (*Source*: Authors' own calculation based on CDEC-SIC 2000, CDEC-SIC 2009, CDEC-SIC 2011) (GWH = gigawatt hour, MW = megawatt)

generation hinges primarily on climate conditions, e.g., on precipitation, air temperature and the storage capacity of mountain regions (soil, underground, snow, glaciers). Precipitation in the MRS, however, is highly variable (Chap. 2), occasionally leading to droughts in the region and consequently deficits in hydropower production (Briggs 2011).

Table 5.1 Electricity generation from thermal power plants in the MRS (*Source*: Authors' own calculation based on CDEC-SIC 2000, CDEC-SIC 2009, CDEC-SIC 2011)

GWh	1990	1992	1994	1996	1998	2000	2002	2004	2006	2008	2010
Nueva Renca	0	0	0	0	2,592	1,841	1,979	2,276	1,484	1,503	1,907
Renca	786	2	31	549	234	14	0	6	0	12	3

5.2.3.2 Generation of Electricity in Thermal Power Stations

Half of the electricity generated in the MRS comes from two thermal power stations, i.e., the *Renca* station, a coal-fired power plant in operation since 1962 (coal has meanwhile been replaced by diesel fuel), and the combined cycle *Nueva Renca* gas power station, which went into operation in 1998 (Table 5.1).

Electricity production in both plants is contingent on SIC market prices. If sufficient low-cost electricity is available from hydropower plants, the thermal power stations do not produce energy. In the case of excess demand, these stations produce electricity, which is then fed into the grid.

The only existing biogas project for electricity generation is at the *Central Loma Los Colorados* I station, which has an installed capacity of two MW. Progressive expansion to reach 28 MW by 2024 is planned (KDM 2010). Although the MRS has significant solar energy potential, photovoltaic installations are so rare that no official record has been produced (interview with López 2010).

5.2.3.3 Heat Generation

LPG and natural gas are used for space heating and/or cooking in 52.4 % of households (houses and apartments) in the MRS, followed by kerosene or paraffin with 30.8 %. An estimated 5.9 % of households uses wood products (e.g., firewood, charcoal, pellets) (CDT 2012). The remaining 11 % uses electricity and city gas (a gas mix consisting of 70 % biogas and 30 % LPG). Water heating in solar collectors is not a common phenomenon. Biogas from *La Farfana* sewage treatment plant distributed through the *Metrogas* network for space heating amounted to merely 0.03 PJ 2008 (Aguas Andinas 2008).

5.3 Future Development of the Energy Sector

In order to estimate the progression of future energy supply and demand in the MRS, energy scenarios were developed for the period up to 2050. The scenarios are based on the "philosophies" for the two framework scenarios "Business as Usual (BAU)" and "Collective Responsibility (CR)" described in Chap. 3 and the basic quantitative variables for these scenarios. Table 5.2 gives an overview of the relationship between the general assumptions made in Chap. 3 and the specific assumptions for the energy sector.

Table 5.2 Energy-related parameters of the two scenario alternatives for the year 2050 (*Source*: Authors' own supplementation based on Barton et al. 2011a, b)

Business as usual (BAU)	Collective responsibility (CR)
The MRS in 2050 is characterized by ongoing recent trends in population, economic growth, urbanization, technology and human behaviour. BAU assumes that current market-based policies will persist and environmental health and ecological integrity be of less interest.	The MRS in 2050 is dominated by strong state presence and market regulation. Environmental protection, social justice and equity are major political goals. Slower economic growth and the introduction of clean and resource efficient technologies are key determinants. CR assumes that population growth in MRS will slow down and decentralization alter urban planning processes.
Urban development	
• Urban population of 8.5 million • Decrease in persons per household as in recent years • Increase in number of households • Increase in net gross built-up living area in line with recent developments • Introduction of new technologies, no focus on energy saving • Tertiarization processes increase along with industrial sector • Real GDP growth in line with historic development but reduced in period from 2030 to 2050	• Urban population of 7.9 million • Slower decrease in persons per household in line with recent development • Deceleration of increase in number of households • Slower increase in net gross built-up living area due to decline in number of households • New technologies include energy-saving and renewable energy technologies • Industrial growth balanced with ecosystems • Real GDP growth is slightly slower than in BAU
Institutional framework	
• Privatization with monopolistic structures • Weak government influence	• Stronger state presence and regulation by government institutions • Promotion of energy-saving technologies
Energy efficiency	
• Increased energy efficiency of electronic devices as in recent years • New building regulations lead to highest increase ever in energy efficiency in buildings • Increased energy efficiency in cars from 2010 to 2030 in line with historic development, with higher values from 2010 to 2050	• State initiatives and changing attitudes lead to use of more energy efficient products • Energy standards for construction increase more than in BAU leading to greater energy efficiency in buildings • Increase in energy efficiency in cars slightly higher than in BAU
Mobility	
• 3.8 million private cars in 2050 • Distance travelled by private passenger cars 20,000 km per car and year in 2010 and 15,000 km per car and year in 2050 • Distance travelled in electric plug-in mode will become more and more common up to 2050, by which time it will have reached 10 %.	• 3.2 million private cars in 2050 • Distance travelled by private passenger cars 20,000 km per car and year in 2010 and 12,000 km per car and year in 2050. • Distance travelled in electric plug-in mode will increase more rapidly and reach 35 % in 2050
Energy infrastructure	
• All hydroelectric stations in 2010, including the new Alto Maipo hydropower plants, still in operation in 2050 • Hydroelectric power plant HydroAisen built and connected to MRS	• In addition to current hydroelectric stations and new Alto Maipo hydropower plants, some small hydroelectric power plants are in operation in 2050

(continued)

Table 5.2 (continued)

Business as usual (BAU)	Collective responsibility (CR)
• Some smaller solar power plants and wind turbines working in 2050 • Renca closed down in 2030, Nueva Renca still in operation • Increased use of biomass gas for heat production	• Hydroelectric power plant HydroAisen has not been built • More solar power plants and wind turbines than in BAU produce electricity in 2050 • Renca closed down in 2030, Nueva Renca operates mainly with gas from renewable resources • Heat production from gas extended and substituted entirely by biomass gas

5.3.1 Future Energy Demand

The demand for energy in the MRS is influenced mainly by demographic development, the number of persons per household, changes in consumption patterns per household, GDP, and technologies that enhance energy efficiency in products, buildings and vehicles. The relationship between socio-economic development and energy consumption was assessed according to the basic philosophies and quantitative variables of the two MRS scenarios BAU and CR (Table 5.2).

Since electricity consumption in the *residential sector* is governed primarily by population size and GDP volume, the growth rate average of the two parameters was interpreted as the future energy growth rate. Expansion of the net floor area in the context of population growth was likewise taken into account since an increase in this indicator leads to a corresponding increase in the use of electricity for lighting and other electronic devices. The results clearly showed a reduction in consumption due to the enhanced energy efficiency of domestic products.

Calculations for the BAU scenario indicate that *electricity consumption* in the residential sector will almost double between 2010 and 2030, and increase by another 20 PJ by 2050. Although the CR scenario also projects an increase in electricity consumption, it is only marginally higher in 2050 than the projection for 2030 in the BAU scenario. This is due to lower population figures, lower net floor area and higher energy efficiency in domestic products (Table 5.3).

Gas and liquid fuels are predominantly used in the residential sector for *heating*. The required amount is therefore correlated to the net floor area to be heated. Increased energy efficiency in building construction minimizes the use of electricity. The consumption of gas and liquid fuels in the BAU scenario in the first 20 years after 2010 will increase by almost 20 %, whereas the increase in the subsequent 20 years averages at only ten per cent. The latter can be explained by the decline in population growth rates, which in turn decelerates the increase in net floor area. CR shows a similar development. Here, however, lower population figures and a higher energy efficiency rates in building construction point to a more controlled progression (Table 5.3).

Energy consumption rates in the *commercial and industrial sectors* are closely linked to GDP growth rates, which slow down as a result of increased energy

Table 5.3 Future energy consumption (*Source*: historical data INE 2001–2007; INE 2010; SEC 2006–2010; future data authors' own calculation)

	History	BAU		CR	
PJ/y	2010	2030	2050	2030	2050
Residential					
Electricity	16.8	33.2	53.2	25.3	35.6
Gas	27.8	33.1	36	30.2	30.9
Liquid fuels	3.8	4.5	4.9	4.1	4.2
Total	48.4	70.8	94.1	59.6	70.7
Commercial					
Electricity	22	59.5	102.7	41.2	56.1
Gas	3.7	9.9	17.2	6.9	9.4
Liquid fuels	1.6	4.4	7.6	3.0	4.1
Total	27.3	73.9	127.4	51.2	69.6
Industry					
Electricity	21.6	46.4	68.6	33.2	41.9
Gas	21.2	45.7	67.5	32.7	41.3
Liquid fuels	3.8	8.3	12.2	5.9	7.5
Total	46.6	100.4	148.3	71.8	90.7
Transport					
Electricity	0.8	3.1	5.8	4.1	13.1
Gas	0.1	0	0	0	0
Liquid fuels	111.9	160.2	143.1	107.1	75.8
Total	112.8	163.3	148.9	111.2	88.9
Total					
Electricity	61.2	142.2	230.3	103.8	146.7
Gas	52.8	88.7	120.7	69.8	81.6
Liquid fuels	121.1	177.4	167.8	120.1	91.6
Total	235.1	408.3	518.8	293.7	319.9

efficiency. Hence lower GDP and higher energy efficiency growth rates in the CR scenario lead to less electricity consumption in CR than in BAU. The increase in consumption in the commercial as distinct from the industrial sector is based on the assumption that future economic activities in the MRS will shift from the industrial to the commercial sector (Table 5.3).

Essential for the calculation of energy consumption in the *traffic sector* is the number of cars owned by MRS inhabitants, the distances travelled and the percentage of these in electric plug-in mode. The latter refers to electricity received from the grid and stored in "on board" batteries rather than produced "on board". This is vital for two reasons: the efficiency of electric vehicles affects the overall consumption of energy in the traffic sector, on the one hand, and leads to an energy shift from liquid fuels to electricity, on the other. Table 5.3 illustrates that—based on historic trends—the consumption of liquid fuels in the BAU scenario will rise. With the increase in energy efficiency and the use of plug-in electric cars, however, liquid fuel consumption will gradually decline and be lower in 2050 than in 2030. Since this shift is expected to appear earlier in the CP scenario and the trend towards electric plug-in vehicles projected to intensify, liquid fuel consumption in 2030 will remain at the level of 2010 in CR and decrease in 2050 by approximately one third. Closely linked to this shift in liquid fuel consumption is the increase in the use of electricity. In 2010, electricity for traffic purposes was consumed solely by the metro. In BAU as of 2030 and in CR as of 2020, however, the increase in the consumption of electricity in the traffic sector will be first and foremost due to an increase in the use of electric vehicles.

Table 5.3 indicates that total energy consumption in the BAU scenario will more than double by 2050, whereas in the CR scenario it is expected to increase by merely 36 %. This outcome for CR is primarily due to a slowing down of GDP and population growth rates in the MRS and greater energy efficiency. Finally, by 2050 the total energy consumption in the CR scenario will correspond to approximately 60 % of the total consumption in the BAU scenario.

5.3.1.1 Future Energy Supply

The assumptions and developments in the CR and BAU framework scenarios indicate an increase in the overall production of electricity within the area of the MRS (Table 5.4). In both scenarios electricity production based on hydroenergy almost matches production in existing hydropower plants and the new hydropower plants in Alto Maipo under construction in 2013. Construction of several small hydropower plants is envisaged between now and 2030 in the CR scenario only. The calculation of future electricity generation is based on historical electricity production, on the one hand, and estimates on the construction of a new electricity production plant as described in the framework scenarios, on the other.

The use of solar (Sánchez et al. 2012; Dietsche 2011; Alvarez 2011) and wind power in CR constitutes a major difference between the two scenarios. The enormous investment in the BAU scenario for construction of the 2,750 MW *HidroAysen* hydroelectric plant and its link to the MRS leaves little or no capital for the construction of further power plants. In the CR scenario, government schemes to stimulate energy production with local resources within and around the MRS reduce investment in new fossil-fired power plants, consequently avoiding high follow-up costs, dependence on imported fossil fuels, and possible protest against the plants.

Inefficiency and high energy production costs lead to the closing down of *Renca* in the 2020s in both scenarios, while *Nueva Renca* continues to play a significant role in electricity production in the MRS, working almost to capacity. Over time, natural gas in the CR scenario will be substituted more and more by methane created from slurry and biomass, and several renewables from other Chilean regions. Biomass is likewise used to feed small decentralized biomass stations erected in a number of MRS municipalities.

Both scenarios consider waste-generated energy, which is gained from the separate collection of landfill gas, the result of the anaerobic break-down of organic waste. The amount of waste to be disposed is reduced, as is the emission of greenhouse gases. This method of energy production is higher in the CR than in the BAU scenario (Bräutigam 2012).

Table 5.4 Future electricity generation (*Source*: Historcial data CDEC-SIC 2010, future data authors' own calculation)

	History	BAU		CR	
PJ/y	2010	2030	2050	2030	2050
Hydroelectric					
Existing	6.1	6.1	6.1	6.1	6.1
Alto Maipo	0.0	9.9	9.9	9.9	9.9
Additional mini-hydro-electric	0.0	0.0	0.0	0.4	0.4
	6.1	16.0	16.0	16.4	16.4
Solar	0.0	0.1	1.5	2.1	10.0
Wind	0.0	0.0	0.0	0.5	2.0
Thermal					
Renca	0,0	0.0	0.0	0.0	0.0
Nueva Renca	6.9	9.0	9.0	9.0	9.0
Decentralized biomass	0.0	0.6	2.2	0.6	3.0
	6.9	9.6	11.2	9.6	12.0
Local electricity	***13.0***	***25.7***	***28.7***	***28.6***	***40.4***

While some 21 % of electricity consumption in the MRS was covered by generation within the region in 2010, the BAU scenario shows a drop in this value to twelve per cent by 2050. In the CR scenario, on the other hand, the same value will rise to 28 % by 2050, rendering the MRS more self-sufficient in terms of energy supplies. In absolute numbers, the electricity production deficit will have increased from 48.2 PJ in 2010 to more than twice that level in CR (106.3 PJ) and four times that level in BAU (201.6 PJ). This bears out the need to diversify the generation of electricity from renewable sources within the MRS and, at the same time, to reduce electricity consumption and increase energy efficiency in order to lessen the dependence on energy imports from outside the region in both scenarios.

5.4 Impact of Climate Change on Future Energy Supply and Demand

Having analysed the energy progression in the BAU and CR scenarios to the exclusion of the adverse effects of climate change, this section now explores the relationship to the latter in more detail. The impact of rising temperatures on heating and cooling was calculated for the demand side based on work by Cai et al. (2011), who developed a method of calculating the consumption of energy in relation to the factors of climate change. As shown in Chap. 2, maximum temperatures in the MRS will rise, while precipitation and stream flow will decline. As precipitation occurs between April and September, most of the water available for the rest of the year relies to a great extent on melt water from glaciers and snowfields in the high Andes. The importance of electricity generation from hydro and thermal power plants for the supply of energy to the MRS calls for closer scrutiny of the impact of climate change on these two technologies.

5.4.1 Impact of Climate Change on Energy Demand

Energy demand will be influenced by climate change in different ways. Temperature increases during the summer will, on the one hand, lead to greater electricity consumption in the residential and commercial sectors to cater for air conditioning and cooling. Forty per cent of electricity consumption in the residential sector is affected by increased temperatures; this figure rises to 55 % in the commercial sector (PRIEN, INAP 2008; ECEE 1999). On the other hand, rising temperatures reduce the consumption of gas and petroleum products for heating in these sectors (notably in the winter). An increase in electricity consumption in the industrial sector for cooling or refrigeration purposes is expected, especially in the food industry.

The following analysis takes temperatures from the A2 climate scenario (Chap. 2) as a reference. Since the difference between the scenarios is negligible, the B1 climate scenario is not taken into account. The annual temperature averages presented in Chap. 2 were calculated as the arithmetic average of maximum and minimum temperatures in order to assess the impact of the expected climate change on energy consumption for heating and cooling. Of the nine existing measuring stations, only the six located no higher than 850 m were considered, since population density beyond this altitude is low. In other words the expected climate change in higher areas will have only a minimal effect on the demand for energy. As Table 5.5 indicates, average annual temperatures will increase by 0.75 °C up to 2030 and by 1.17 °C up to 2050.

Cai et al. (2011) introduced a method of calculating energy consumption in relation to average temperature development. It incorporates multiple technologies, resources, sectors and climate change impacts into a general model, where five fuzzy sets represent the different impact levels of climate change on the demand for energy. In the demand analysis, these fuzzy sets are sensitive to variations in climatic conditions, particularly those related to heating and cooling in the residential and commercial sectors.

The adverse effects of climate change on the demand for energy are categorized in a hierarchy of five levels (neutral, low, medium, high, very high) represented as five fuzzy sets. It was assumed that the energy demand would increase by one to five per cent compared to the original amount. Temperature levels were taken from results obtained by Cortés et al. (2012).

Using this methodology, energy consumption will increase by 0.25 % when temperatures increase by 0.1°C. Hence a rise in temperature of 0.75 °C between 2010 and 2030 will lead to an increase in energy consumption of 1.88 % and of another 1.05 % by 2050 due to a further rise in temperature of 0.42°C. The increase in energy consumption between 2010 and 2050 will amount to 2.93 %.

The increase in energy consumption for climatization and refrigeration purposes in the different sectors does not coincide with increased energy for lighting and electronic devices, as these are not affected by increased temperatures.

In the *residential sector*, climatization (PRIEN-INAP 2008; ECEE 1999) contributes with 9 % and refrigeration with 31 % to the total consumption of

Table 5.5 Temperature averages in the MRS in 2030 and 2050 based on the A2 climate scenario for the locations Cerro Calan, Florida, Melipilla, Pirque, Santiago and Tobalaba (*Source*: Authors' own presentation based on Cortés et al. 2012)

Period	Historic average temperature (1970–2010)	A2 climate scenario 2030	A2 climate scenario 2050
Temperature averages (°C)	15.34	16.09	16.51
Difference from historic average (°C)		0.75	1.17

electricity. As a result of climate change, this will lead to a 0.75 % increase in consumption by 2030 and a 1.17 % increase by 2050. By 2050, therefore, the residential sector will see an increase in electricity consumption from 53.2 PJ excluding climate change to 53.8 PJ including climate change in the BAU scenario and from 35.6 PJ excluding climate change to 36.0 PJ including climate change in the CR scenario (Table 5.6).

In the *commercial sector*, climatization contributes 20 % and refrigeration 35 % to the total consumption of electricity (PRIEN-INAP 2008; ECEE 1999). Accordingly, climate change will lead to an increase in consumption of 1.03 % by 2030 and 1.61 % by 2050, which translates to an increase from 102.7 PJ to 104.4 PJ in BAU and 56.1 PJ to 57.0 PJ in CR by 2050 (Table 5.6).

In the *industrial sector*, electricity consumption for air conditioning and refrigeration purposes constitutes the lowest share among the three sectors (three per cent, Gama Ingenieros 2009; PROCOBRE 1994). An additional 0.06 % electricity will be consumed in this sector by 2030, and 0.09 % by 2050. The figures are infinitesimal and visible only in the Bau scenario as a shift from 68.6 to 68.7 (Table 5.6).

The methodology of Cai et al. (2011) was also used to calculate the effect of climate change on the *use of gas*. Gas consumption drops by 0.25 % with every 0.1°C rise in temperature. In the residential sector, 50 % of the gas is used for space heating. Rising temperatures, however, will lead to a decline in the use of gas in this sector of 0.9 % by 2030 and 1.5 % by 2050, or a drop from 36.0 to 35.5 PJ by 2050. Ninety per cent of the gas consumption in the commercial sector is used for heating; here there will be a reduction of 2.6 % by 2050. The overall effect in the BAU scenario is a reduction in gas consumption from 17.2 to 16.7 PJ and in the CR scenario from 9.4 to 9.2 PJ.

Electricity consumption in the BAU scenario will increase from 230.3 PJ to 232.6 PJ and in the CR scenario from 146.7 to 148.1 PJ. Gas consumption, on the other hand, will be reduced in BAU from 120.7 to 119.7 and in CR from 81.6 to 80.9 PJ. In total, climate change leads in BAU to an increase in energy consumption from 518.8 to 520.2 PJ by 2050 and in CR from 319.9 to 320.6 PJ.

In summary it can be said that of the four sectors discussed, the residential and commercial sectors are affected most, albeit climate change has very little impact on energy consumption as a whole.

Table 5.6 Future energy consumption including climate change (*Source*: Historical data INE 2001–2007, INE 2010, SEC 2006–2010, future data authors' own calculation, CC influence calculated due to Cai et al. 2011)

	History	BAU		CR	
PJ/y	2010	2030	2050	2030	2050
Residential					
Electricity	16.8	33.2	53.2	25.3	35.6
incl. CC		**33.4**	**53.8**	**25.5**	**36.0**
Gas	27.8	33.1	36.0	30.2	30.9
incl. CC		**32.8**	**35.5**	**29.9**	**30.4**
Liquid fuels	3.8	4.5	4.9	4.1	4.2
Total without CC	48.4	70.8	94.1	59.6	70.7
Total with CC		**70.7**	**94.2**	**59.5**	**70.7**
Commercial					
Electricity	22	59.5	102.7	41.2	56.1
incl. CC		**60.1**	**104.4**	**41.6**	**57.0**
Gas	3.7	9.9	17.2	6.9	9.4
incl. CC		**9.7**	**16.7**	**6.8**	**9.2**
Liquid fuels	1.6	4.4	7.6	3.0	4.1
Total without CC	27.3	73.9	127.4	51.2	69.6
Total with CC		**74.2**	**128.7**	**51.4**	**70.3**
Industrial					
Electricity	21.6	46.4	68.6	33.2	41.9
incl. CC		**46.4**	**68.7**	**33.2**	**41.9**
Gas	21.2	45.7	67.5	32.7	41.3
Liquid fuels	3.8	8.3	12.2	5.9	7.5
Total without CC	46.6	100.4	148.3	71.8	90.7
Total with CC		**100.4**	**148.4**	**71.8**	**90.7**
Transport					
Electricity	0.8	3.1	5.8	4.1	13.1
Gas	0.1	0	0	0	0
Liquid fuels	111.9	160.2	143.1	107.1	75.8
Total without CC	112.8	163.3	148.9	111.2	88.9
Total					
Electricity	61.2	142.2	230.3	103.8	146.7
Electricity incl. CC		**143.1**	**232.6**	**104.4**	**148.1**
Gas	52.8	88.7	120.7	69.8	81.6
Gas incl. CC		**88.2**	**119.7**	**69.4**	**80.9**
Liquid fuels	121.1	177.4	167.8	120.1	91.6
Total without CC	235.1	408.3	518.8	293.7	319.9
Total with CC		**408.7**	**520.2**	**293.9**	**320.6**
CC influence		**+0.4**	**+1.4**	**+0.2**	**+0.7**

5.4.2 Impact of Climate Change on Energy Supply

To gain an insight into the effects of climate change on the supply of energy, this section discusses: (1) the impact of decreased stream flow (Chap. 2) on hydroelectricity production and (2) the impact of rising temperatures on thermal power stations.

5.4.2.1 Future Electricity Generation in Hydroelectric Power Stations

Electricity production in hydroelectric power stations is directly proportional to the amount of water (in m^3) used by the turbine and the penstock. The amount of water was calculated with monthly stream flow data from the Maipo river. As each turbine has a specific maximum capacity, only the amount below the maximum can be used for electricity production. Hence stream flows above this level do not translate to higher electricity production. Stream flow data above the maximum in question were therefore set at this maximum. Table 5.7 compares "uncorrected" and "corrected" stream flow data for the 1996–2010 period and the two climate scenarios, A and B (both scenarios are taken into account in case major differences arise in future energy production). In this case, a stream flow of 100 m^3/s was used for Alfalfal, the hydroelectric plant with the highest capacity. The decrease in average monthly stream flow by 2050 is much lower when the corrected data relevant to electricity production is taken into account (five per cent lower in scenario A2, and eight per cent lower in scenario B1).

Based on stream flow data measurements at the Maipo San Alfonso station (average monthly values cited in Cortés et al. 2012) and the corresponding electricity production values at the hydropower stations referenced in Fig. 5.3, a linear correlation function (electricity production as a function of stream flow) was calculated for each hydropower station.

The correlation functions were then applied to the calculation of monthly values for electricity production for the years 2020–2059, based on monthly stream flow data for the two climate scenarios A2 and B1 (Chap. 2) cited in Cortés et al. (2012). Electricity production for the reference year 2030 was calculated as an annual average value over the calculated monthly values for the years 2020–2039 and for the reference year 2050 as an annual average value over the monthly values for the years 2040–2059. Table 5.8 shows the results.

In the A2 scenario, electricity production increases from 1,736 GWh in the 1996–2010 period to 1,901 GWh in 2030 (a 9.5 % increase). Electricity production will reach 1,681 GWh by 2050, a decrease of 3.2 % compared to the 1996–2010 period. Electricity production is lower in the B1 than in the A2 scenario. From 1,736 GWh in the 1996–2010 period, it will reach 1,681 GWh (a 2.7 % increase) by 2030 and decrease to 1,639 GWh by 2050 (a 5.6 % decrease compared to the 1996–2010 period) (Table 5.9). The A2 climate scenario, which is more critical than the B1 scenario, was used to calculate potential climate change effects in the BAU and CR scenarios.

The resultant calculations show no increase (or decrease) in the frequency of low electricity production periods due to extremely low stream flows (in dry periods). This implies that there will be no major alteration in hydroelectricity production in existing hydropower plants due to climate change up to 2050. Nevertheless, hydroelectric power plants in the MRS under construction in the Alto Maipo (Alfalfal II and Las Lajas) hydropower project may be more sensitive to future run-off reductions. This would increase the existing risk of electricity shortages due to run-off reductions resulting from climate change.

Table 5.7 Monthly average stream flow data for different time periods and different scenarios for Maipo San Alfonso (*Source*: Authors' own presentation based on Cortés et al. 2012)

Time period and scenario	Stream flow (m^3/s) (monthly averages) ("uncorrected")	Stream flow (m^3/s) values above 100 m^3/s are set to 100 m^3/s (monthly averages) ("corrected")
1996–2004	78.3	61.7
2020–2039 (A2)	87.1	70.3
2040–2059 (A2)	63.5	58.7
2020–2039 (B1)	75.0	64.0
2040–2059 (B1)	63.1	56.8

Table 5.8 Electricity production by power stations in different time periods and different scenarios (*Source*: Authors' own calculation based on CDEC-SIC 2009)

Hydropower stations	Average electricity production 1996–2010	Calculated electricity production			
		2020–2039	2040–2060	2020–2039	2040–2059
		A2 scenario		B1 scenario	
	Annual average in GWh				
Alfalfal	870	1 026	857	935	829
Queltehues	361	388	372	378	365
Maitenes	130	128	125	126	124
Florida	148	143	122	132	119
Puntilla	122	108	102	105	101
Volcan	105	108	102	105	101
Total	**1 736**	**1 901**	**1 681**	**1 782**	**1 639**

5.4.2.2 Future Electricity Generation in Thermal Power Stations

To evaluate the impact of rising temperatures on electricity generation in the combined cycle Nueva Renca thermal power station, the largest plant in the MRS, production levels at the gas turbine were analysed in light of the fact that the ambient conditions of air entering the turbine impacted heavily on its productivity and efficiency.

Combustion gas turbine capacity rates are generally based on standard ambient air conditions as specified by the International Organization for Standardization (ISO) (ASHRAE 2008). The corresponding air inlet conditions are: air temperature 15°C; relative humidity 60 %; absolute pressure 101.3 kilopascal (kPa) at sea level.

Thermodynamic analyses taken from the literature indicate that thermal efficiency and specific output decrease in correlation to the increase in humidity and ambient temperature. Ambient temperature is the variable with the greatest impact on gas turbine performance (Farzaneh-Gord and Deymi-Dashtebayaz 2011). This is because a rise in temperature leads to a decrease in air density and, consequently, to

Table 5.9 Future electricity supply with climate change included (*Source*: Historical data CDEC-SIC 2012, future data authors' own calculation)

	History	BAU		CR	
PJ/y	2010	2030	2050	2030	2050
Hydroelectric					
Existing	6.1	6.1	6.1	6.1	6.1
Alto Maipo	0.0	9.9	9.9	9.9	9.9
Additional mini-hydro-electric	0.0	0.0	0.0	0.4	0.4
Subtotal	6,1	16.0	160	16.4	16.4
incl CC		**16.4**	**15.1**	**16.8**	**15.5**
Solar PV	0,0	0.1	1.5	2.0	10.0
Wind	0,0	0.0	0.0	0.5	2.0
Thermal					
Renca	0.0	0.0	0.0	0.0	0.0
Nueva Renca	6.9	9.0	9.0	9.0	9.0
Decentralized biomass	0.0	0.6	2.2	0.6	3.0
Subtotal	6.9	9.6	11.2	9.6	12.0
incl CC		**9.6**	**11.1**	**9.6**	**11.9**
Local electricity	15.7	25.7	28.7	28.6	40.4
incl. CC		**26.1**	**27.7**	**28.9**	**39.4**
CC effect		**+0.4**	**−1.0**	**+0.3**	**−1.0**

a reduction in the mass flow rate. Less air passes through the turbine and power output is reduced. Compression work also increases as a result of low air density.

In their work, Ibrahim, Rahman, and Abdalla (2011) show that an increase of one degree C in compressor air inlet temperature reduces gas turbine power output by one per cent. Moreover, peak periods for electricity demand occur in the summer when ambient temperatures are higher. Other authors like Valor et al. (2001), Mohanty and Paloso (1995) and Kakaras et al. (2004) analyse the relation between electricity losses and air temperature. They conclude that productivity decreases as the inlet air temperature in the turbine increases.

The effect on the Renca gas turbine was calculated by analysing the behaviour of variables in a Siemens V84.3 turbine. To assess the impact of climate change, the percentage of nominal power and lost productivity was determined by varying the temperature of the inlet air; the average temperature for the A2 climatic scenario was taken into account (Table 5.8). The results are shown in Table 5.10.

Average temperatures in the MRS show that the maximum change in temperature by 2030 will be 0.75 °C and 1.17 °C by 2050. This means that by 2030, the electricity generating capacity of the turbine will drop by 807 kW from an initial nominal capacity of 191,492 kW. By 2050, the capacity will decrease by another 461 kW.

This reduction in efficiency has two effects. More gas will be required to produce the same amount of electricity unless the turbine works to capacity, making production more expensive. Moreover, when the turbine performs to maximum

Table 5.10 Impact of temperature on nominal power of the gas turbine (*Source*: Authors' own calculation based on SIEMENS 2004; Cortés et al. 2012)

Period	Average	1970–2010	2030–2050
Average temperature (°C)	15.4	16.1	16.5
Cumulated temp. difference (°C)	0	0.75	1.17
Reduction (kW)	0.	807	1,268
Nominal power (kW)	191,492	190,686	190,224
Loss of efficiency (%)	0	0.4	0.7

capacity it produces less power. Consequently, the energy produced will decrease by 0.4 % by 2030 and 0.7 % by 2050. These percentages are so small as to be nearly invisible in the calculations in Table 5.9.

5.4.3 Future Energy Situation Including Climate Change

The differences between the two scenarios attest to the complexity of the future energy framework of the MRS, since it depends on factors such as growth rates in population and GDP, political contributions to the production of renewable energy, and investment in HidroAysen. The impact of climate change on energy supply and demand is negligible (Tables 5.6 and 5.10). Projected future temperature rises in each of the scenarios lead to a minor increase in the total demand for energy. On the other hand, there will be a slight reduction in hydroelectricity production and production efficiency in thermal power plants. The risk of lower energy generation from hydropower in dry years, on the other hand, will most likely remain at its current level. This increases the risk of failures in the electricity production system in the MRS and its dependency on energy imports.

5.5 Measures

A review of the literature was undertaken to find out the most effective strategies and measures to reduce the risk of future problems in the MRS energy system. Most of the cities with successful risk-reduction methods in the energy system were found in Europe (Covenant of Mayors 2012; EEA 2012; ICLEI 2012). This theoretical research was complemented by a series of interviews and a workshop with stakeholders from the energy sector in the MRS, and resulted in the recommendation of the following strategic approaches in order of priority:

1. Diversification of energy sources in electricity generation, primarily by reducing the high dependency on hydropower and thermal power.
2. Limitation (and ultimately reduction) of energy consumption growth rates in order to reduce future vulnerability of the power supply in the MRS.

Implementing these basic energy policy strategies requires suitable political and institutional framework conditions, the availability and use of appropriate technologies and products, and, last but not least, both public and company awareness and support. Successful policy implementation is currently obstructed by a number of difficulties in the MRS, which need urgent attention:

- Lack of information, sensitivity and awareness among economic and political decision-makers, the administration and the general public with regard to current and future risks associated with the supply of energy and their link to the effects of climate change.
- Lack of experts and expert knowledge to advise the regional and local administrations in matters of energy efficiency and energy supply diversification that would allow them to adequately support and accompany these strategies and enhance their implementation in the context of national energy policy decision-making in BAU, but also in CR.
- Independent actions by political and sectorial decision-makers at different regional levels to promote diversification of the energy supply. One case in point is the decision to build a system of dams in the Aysen region to supply the MRS with a major percentage of its electricity needs. On the one hand, this leads to stability of the energy supply in the MRS but, on the other hand, makes the long power line connecting the production site in Aysen to the MRS extremely vulnerable to volcanos, earthquakes, bush fires and terrorism. On the whole, this undertaking would render the energy supply of the MRS more vulnerable than a decentralized energy production system in closer proximity.
- Lack of suitable incentives to exploit existing technologies that would enhance energy efficiency in, for instance, building construction and renovation.
- To overcome these obstacles and implement a sound energy policy in the MRS, a set of 36 measures was proposed and listed in the Regional Climate Change Adaptation Plan for Santiago de Chile (cf. Chap. 9). The list was based on literature reviews, experience in other cities and numerous discussions with Chilean experts and regional stakeholders. Finally, four measures or groups of measures were selected according to strategic importance, effectiveness and broad social acceptance. Following scientific debate and local expertise, and considering the relevance of the transport sector for the supply of energy, albeit not a focus of this study, the measures focused on three elements:
  1. Modification of institutionalized settings
  2. Improvement of energy-related education and information
  3. Improvement of incentives to achieve maximum energy efficiency in buildings. These measures are addressed to the government and municipalities of the MRS and to Chilean government ministries. In the following the four measures are briefly presented.

5.5.1 Setting up Energy Expert Groups Within the Regional Government and the Municipalities

The first measure sees the establishment of energy expert groups within the Regional Government and the municipalities of the MRS, composed of scientific and technical experts on energy matters. The experts would act in an advisory capacity to decision-makers, private households, companies and commerce at metropolitan and municipality levels. Moreover they could help to define modified future energy policy guidelines, design public and private programmes and initiatives to promote renewable energies and energy efficiency improvements, and ensure more effective use of public and private financial resources. For this purpose, the groups should have sufficient knowledge of regional potentials, and of the technical and organizational preconditions for exploitation of local renewable energies and the increase of energy efficiency in the different contexts. The service should be free of charge or priced according to customer income.

In the medium term, the group attached to the Regional Government of the Metropolitan Region of Santiago de Chile (GORE) should consist of approximately ten experts, complemented by one expert per 100,000 inhabitants in each of the MRS municipalities.

The expert groups should be responsible for public campaigns to promote and raise awareness of the diversification of energy production with more extensive use of local renewable energy sources, and of the need for energy efficiency measures.

GORE and the municipality expert groups should work closely with key players in the energy system, such as the national Ministry of Energy, Regional Council of Chile (CORE), Chilean Energy Efficiency Association (AChEE), the Ministry of Environment, National Program of Energy Efficiency (PPEE), Centre for Renewable Energy (CER), Chilean Renewable Energy Association (ACERA), Chilean Corporation for Production Promotion (CORFO), and energy utilities such as Metrogas, Chilectra.

5.5.2 Education and Information on Climate Change and Energy-Related Issues

The basic idea of this second measure is to improve methods of educating and informing the general public in the MRS about the fundamentals of climate change, the expected consequences for the MRS in general and for the energy sector in particular, and of adapting the energy sector to increase energy efficiency and exploit renewable energy. This could help to reduce resistance to envisaged changes in the system or distrust of new technologies to be used in adaptation measures.

Knowledge and information to this effect should be disseminated in institutions of primary, secondary and tertiary education, and in adult education centres. This could be designed, organized and implemented to a certain extent in public–private partnerships with the business sector. Education and training courses would be conducted by various institutions of education or technical training organizations authorized by National Service for Training and Profession (SENCE) (OTEC, Technical Training Agency), which would design a curriculum and seek approval of the Ministry of Education in cooperation with AChEE.

5.5.3 Diversification of Energy Sources in Energy Production

The idea behind the third measure is to make use of more diverse, notably local, renewable energy sources and reduce the vulnerability of the MRS electricity supply, including the risk of interruption of the transmission line that connects the designated HidroAysén hydropower plant with the MRS. Increased exploitation of local renewable resources would require some regulatory changes in the power transmission system (e.g., preference for local renewables to feed electricity into the power grid system, connection of the SIC to the Central Interconnected System (SING) grid) and in the distribution system (e.g., allow for connections to be made to "small and medium-size electricity generation units", implementation of smart grids and of measures to establish more competition at final customer tariff level). The Ministry of Energy, CNE, SEC, the Ministry of Environment and CER / AChEE would be responsible for initiating the process towards such activities.

5.5.4 Reduction of Energy Consumption in Buildings

The fourth measure aims at reducing energy consumption, since the stock of residential and public and commercial buildings is responsible to 19 and 11 %, respectively, for energy consumption in the MRS, primarily for space heating and cooling. Recent decades have seen the increase on the market in Europe and the United States of innovative buildings that consume less energy, including so-called "zero-net-energy buildings" and "passive houses". In general, building sectors in cities have tremendous potential to reduce the consumption of energy.

Against this background, three steps are proposed for the MRS with respect to existing buildings and the design of new structures:

- The public sector could begin to refurbish its buildings energetically and construct new "zero-net-energy buildings" as forerunners to show that this strategy is both economically and environmentally valid for the MRS.
- Establishment of a public economic incentive programme that would provide funds (e.g., subsidies or loans at low interest rates) for energy efficiency-related renewal, construction of "zero-net-energy buildings" or training of players involved in the construction sector (e.g., architects, craftsmen, planners).
- Integration of specific standards (e.g., the "zero-net-energy building") as legal norms into the code for new or existing buildings.

Proper implementation of these measures clearly calls for concerted action at national, regional and local levels.

5.6 Conclusions

The results of the analyses outlined in this chapter show first of all that climate change is unlikely to have a substantial impact on the energy sector in the MRS. Nevertheless, the heavy increase in the demand for energy in the future as a result of population and GDP growth rates will lead to a greater risk of power failure or growing dependence on energy imports if changes are not brought about in the MRS/Chilean energy system.

Two strategies to deal with these risks were proposed: diversification of energy sources for power production via increased use of local renewable energy resources and less energy consumption as a result of more energy efficiency.

Proper implementation of these strategies demands coordinated political and administrative measures that involve the local population. Decision-makers in the MRS and in Chile as a whole could learn from cities that have already implemented these strategies successfully, e.g., Copenhagen in Denmark, Graz in Austria or Freiburg in Germany, all of which are located in regions with far less potential to exploit local renewable energy than the MRS. The crucial task here is to analyse the extent to which these approaches and the experience of establishing proper political and institutionalized framework conditions, and solving organizational issues (e.g., municipal energy utilities) in financial or participatory schemes can be transferred to the Chilean/MRS case and adapted to conditions there.

References

Aguas, A. (2008). Sustainability report of Aguas Andinas 2008. p. 113. Santiago.

Alvarez, F. (2011). Disponibilidad de EERR en Chile. Santiago.

ASHRAE – American Society of Heating, Refrigerating and Air-Conditioning Engineers. ASHRAE Handbook – HVAC Systems and equipment (SI). Atlanta.

Barton, J., Kopfmüller, J., Stelzer, V. (2011a). Project "ClimateAdaptationSantiago" (CAS): Quantitative variables: tendencies and scenario estimations 2010–2050, Working Paper.

Barton, J., Kopfmüller, J., Stelzer, V. (2011b). Project "ClimateAdaptationSantiago" (CAS): Scenario storylines for the RMS, Working Paper.

Bräutigam, K.-R. (2012). Contribution of an optimized waste management to energy supply in santiago de chile under climate change conditions. The ISWA Solid Waste Congress 2012, Florence, Italy, 17–19 September, http://www.iswa.org/en/525/knowledge_base.html?tx_iswaknowledgebase_searchbox[searchphrase]=klaus-rainer&cHash=2e7d11d334c6360f8a957f890b322915. Accessed August 2012.

Briggs, M. (2011). Chile's reliance on hydropower: a risky gamble in an increasingly dry energy climate. In: The Santiago Times, 20 February 2011. http://www.santiagotimes.cl/opinion/op-ed/20800-chiles-reliance-on-hydropower-a-risky-gamble-in-an-increasingly-dry-energy-climate. Accessed March 2013.

Cai, Y. P., Huang, G. H., Tan, Q., & Yang, Z. F. (2011). An integrated approach for climate change impact analysis and adaptation planning under multi-level uncertainties. Part I: methodology. *Renewable and Sustainable Energy Reviews, 15*(2011), 2779–2790.

CER – Center of Renewable Energy (2012). Report CER March. http://cer.gob.cl/boletin/marzo2012/boletin_CER_marzo.pdf. Accessed September 2012.

Central Energía – Central de información y discusión de energía en Chile (2012). Centrales. http://www.centralenergia.cl/centrales/. Accessed March 2013.

Cortés, G., Schaller, S., Rojas, M., Garcia, L., Descalzi, A., Vargas, L., & McPhee, J. (2012). *Assessment of the current climate and expected climate changes in the Metropolitan Region of Santiago de Chile*. Leipzig: UFZ Report.

Covenant of Mayors (2012). http://www.eumayors.eu/index_en.html, Accessed November 2012.

BCC – Banco Central de Chile (2011). PIB Regional 2011. Excel database. http://www.bcentral.cl/estadisticas-economicas/publicaciones-estadisticas/index.htm. Accessed September 2012.

CDEC-SIC – Centro de Despacho Económico de Carga del Sistema Interconectado Central (2000). Operation statistics years 1990–1999, p. 24. https://www.cdec-sic.cl/contenido_en.php?categoria_id=4&contenido_id=000034. Accessed August 2012.
CDEC-SIC – Centro de Despacho Económico de Carga del Sistema Interconectado Central (2009). Yearbook 2009. Operation statistics years 1999–2008, p. 23. http://www.cdec-sic.cl/datos/anuario2009/cdecing/index_ing.htm. Accessed August 2012.
CDEC-SIC – Centro de Despacho Económico de Carga del Sistema Interconectado Central (2011). Yearbook 2010. Operation statistics years 2001–2010. http://www.cdec-sic.cl/datos/anuario2011/ingles/index.html. Accessed August 2012.
CDEC-SIC – Centro de Despacho Económico de Carga del Sistema Interconectado Central (2012). Yearbook 2011. Operation statistics years 2002–2011. https://www.cdec-sic.cl/contenido_en.php?categoria_id=4&contenido_id=000034 Accessed August 2012.
Dietsche, J. (2011). Entwicklungsperspektiven der Photovoltaik in Chile. Karlsruhe.
ECEE – Export Council for Energy Efficiency (1999). Market assessment of Chile – electricity end use and efficiency potential. http://www.ecee.org/pubs/assess/chile/Chile4.htm. Accessed September 2012.
EEA – (2012) European Energy Award (2012). http://www.european-energy-award.org/index.php?id.Accessed November 2012.
Farzaneh-Gord, M., & Deymi-Dashtebayaz, M. (2011). Effect of various inlet air cooling methods on gas turbine performance. *Energy, 36*, 1196–1205.
Gama Ingenieros, S.A. (2009). Final Report to CNE – PPEE – ASIQUIM. Caracterización del consumo energético en la industria quimica chilena.
Gas Natural Quintero (2012). Homepage Company of natural gas liquefied, Quintero. Available in: http://www.gnlquintero.cl/nosotros.htm. Accessed December 2012.
Ibrahim, T. K., Rahman, M. M., & Abdalla, A. N. (2011). Improvement of gas turbine performance based on inlet air cooling systems: A technical review. *International Journal of Physical Sciences, 6*(4), 620–627.
ICLEI – Local governments for sustainability (2012). http://www.iclei.org/. Accessed November 2012.
IEA – International Energy Agency. (2011). *CO2 emission from fuel combustion. IEA*. Paris: OECD.
INE – National Institute of Statistics (2001–2007). Subdepartamento Estadisticas de Transporte y Comunicaciones. Parque de vehiculos en circulacion, vol. 2001 a 2007. http://www.ine.cl/canales/chile_estadistico/estadisticas_economicas/transporte_y_comunicaciones/parquevehiculos.php. Accessed August 2012.
INE – National Institute of Statistics (2010). Distribución de energía eléctrica según región y tipo de cliente 2007–2010. http://www.ine.cl/canales/chile_estadistico/estadisticas_economicas/energia/series_estadisticas/informacion_2010.php. Accessed August 2012.
Kakaras, E., Doukelis, A., & Karellas, S. (2004). Compressor intake-air cooling in gas turbine plants. *Energy Journal, 29*, 2347–2358.
KDM – kdm energía s.a. (2010). La experiencia del relleno sanitario Loma Los Colorados en captura de biogas y Generacion de energia en el marco de proyectos MDL. http://www.latincarbon.com/2010/docs/presentations/Day2/Sergio_Durandeau.pdf, Accessed September 2012.
López, G. (2010) (Entrevista, 2010). correspondencia personal con Gabriela López, ACESOL (Chilean solar energy association) executive assistant, Santiago de Chile, 15 April 2010.
Mansur, E. T., Mendelsohn, R., & Morrison, W. (2008). Climate change adaptation: A study of fuel choice and consumption in the US energy sector. *Journal of Environmental Economics and Management, 55*(2008), 175–193.
ME – Energy Ministry (2010): National Balance of Energy from 1991 to 2010. Table 2 – secondary energy balance. http://antiguo.minenergia.cl/minwww/opencms/14_portal_informacion/06_Estadisticas/Balances_Energ.html. Accessed August 2012.
Ministerio de Planificación (2006). Encuesta Casen 2006 – Energía. Santiago.

Mirasgedis, S., Sarafidis, Y., Georgopoulou, E., Kotroni, V., & Lagouvardis, D. P. (2007). Modeling framework for estimating impacts of climate change on electricity demand at regional level: Case Greece. *Energy Conversion and Management, 48*(2007), 1737–1750.

Mohanty, B. G., & Paloso, J. R. (1995). Enhancing gas turbine performance by intake air cooling using an absorption chiller heat recovery systems. *CHP, 15*, 41–50.

PRIEN-INAP (2008) – Program of studies and investigation in energy – INAP - Institute of Public Affairs at the University of Chile. Estimacion preliminar del potencial de la eficiencia en ele uso de la energia electrica al abastecimiento del sistema interconectado central - SIC. P. 30

PROCOBRE (1994). Uso eficiente de la energía eléctrica. Procobre y Gobierno de Chile.

Sánchez, R., Marchant, C., & Borsdorf, A. (2012). The role of Chilean mountain areas in time of drought and energy crises: new pressures and challenges for vulnerable ecosystems. *Journal of Mountain Science, 9*(2012), 451–462.

SEC – Chilean Superintendence of Electricity and Fuels (2006 - 2010): Statistical report of production, import and sale of crude oil, natural gas and derivatives, vol. 2006 - 2010. http://www.sec.cl/portal/page?_pageid=33,3429539&_dad=portal&_schema=PORTAL. Accessed August 2012.

SIEMENS (2004). Technical papers of gas turbine Siemens V 84.3. http://www.energy.siemens.com/hq/pool/hq/energy-topics/pdfs/en/gas-turbines-power-plants/2_Technology_Evolution.pdf. Accessed August 2012.

Simon, S., Stelzer, V., Vargas, L., Paredes, G., Quintero, A., & Kopfmüller, J. (2012). Energy system. In D. Heinrichs, K. Krellenberg, B. Hansjürgens, & F. Martínez (Eds.), *Risk Habitat Megacity* (pp. 183–205). New York: Springer.

Valor, E., Meneu, V., & Caselles, V. (2001). Daily air temperature and electricity load in Spain. *American Meteorological Society, 2001*, 1413–1421.

The Impacts of Climate and Land-Use Change on Flood and Heat Hazards

6

Annemarie Müller and René Höfer

Abstract

Urban growth and climate change are the primary causes of hydro-meteorological hazard generation in cities. This contribution takes the Metropolitan Region of Santiago de Chile (MRS) as an example of how land-use change has influenced flood and heat hazards and the exposure of built-up areas to both phenomena. It applies remote sensing, GIS, hydro-meteorological and census data to derive quantitative findings on the impact of land-use and climate change on flood and heat hazards. The analysis clearly proves that flood and heat hazard generation is not determined by climate changes alone but also by the shift in urban land-use patterns. Explorative scenarios that describe the variables most relevant to hazard generation are analysed to gain insight into the future development of both extreme events in the MRS. Results show that despite the different intensities of the scenario alternatives, flood and heat hazards will increase in the future, calling for specific adaptation measures to counter both phenomena.

Keywords

Land-use change • Generation of flood and heat hazard • Climate change adaptation

A. Müller (✉)
Department of Urban and Environmental Sociology, Helmholtz Centre for Environmental Research – UFZ, Permoser Str. 15, 04318 Leipzig, Germany
e-mail: annemarie.mueller@ufz.de

R. Höfer
Department Groundwater Remediation, Helmholtz Centre for Environmental Research – UFZ, Permoser Str. 15, 04318 Leipzig, Germany

K. Krellenberg and B. Hansjürgens (eds.), *Climate Adaptation Santiago*,
DOI 10.1007/978-3-642-39103-3_6, © Springer-Verlag Berlin Heidelberg 2014

6.1 Introduction

Climate change, which is predicted to intensify in the future (cf. Chap. 2), and land-use change, mainly caused by ongoing urban expansion, have a significant influence on the occurrence of flood and heat hazard in the Metropolitan Region of Santiago de Chile (MRS). The high level of spatial expansion in the past resulted in added sealed surfaces and the loss of valuable vegetated retention and cooling areas, both in the central built-up body and on the outskirts. This led to disturbance of the MRS ecosystem and to an increase in flood and heat hazards (Romero et al. 1999; Romero and Vásquez 2005). Detailed studies of previous land-use developments and their impact on both hazards have yet to be undertaken for the entire MRS. Moreover, the projected influence of climate change on the occurrence of both meteorologically extreme events has not been investigated so far.

Loss of green spaces and agriculturally used land to new residential sites and commercial centres reduces the availability of retention areas to manage surface run-off during and after precipitation events. More extensive surface sealing lessens the soil infiltration capacity, resulting in an increase in flood hazard. In addition to that vegetation loss leads to higher air temperatures, since the valuable cooling function of vegetation through transpiration processes is reduced considerably. This in turn magnifies heat hazard and underlines the impact of urban expansion on the local climate.

In addition, climatic factors influence both hazards. With more than 50 days of maximum temperatures over 30 °C in the central parts of the MRS (Cortés et al. 2012), heat hazards are frequent in the summer in many MRS areas. Heat hazard occurs in areas where the surface temperature is a standard deviation above the average surface temperature of built-up areas. Although average rainfall is comparatively low (currently 332.3 mm in the central areas of the MRS and 442.9 mm in the eastern part of the city at a height of 920 meter above sea level (masl) (own calculations of rainfall statistics, DGA)), floods occur almost bi-annually during the winter months, as the MRS is ill-equipped to dispose of the vast amounts of storm water following intense rainfall events. The flood height depends on the local relief, but seldom exceeds 20 cm. It frequently interrupts urban functions, however, and is harmful in one way or another to vulnerable households (Müller et al. 2011). The climate will be drier in the future (cf. Chap. 2), increasing heat hazard and the loss of natural vegetation along the urban fringe. This leads to depletion of the retention areas along the urban buffer zone and consequently to an upsurge in flood hazard.

Hence both hazards are currently influenced by urban expansion and the prevailing climatic conditions. In addition, exposure to these hazards is growing as more people are settled in hazard-prone areas and hazard zones expand.

This study shows the land-use changes derived from satellite data between 2001 and 2009, and relates them to changes in flood and heat hazard and the exposure of built-up areas to both hazards. In addition it explores the correlation between land-use changes and a changing climate with reference to the distribution of hazard zones in the MRS. The leading research questions are:

- What is the interrelationship between changes in land-use/land-cover (LULC) and changes in flood and heat hazard?
- What is the future impact of climate and land-use change on flood and heat hazard?

In response to these research questions the study combines quantitative and qualitative methods and a variety of data sets (Sect. 6.2). Past land-use changes in the MRS are analysed in relation to the distribution of both hazard zones (Sect. 6.3). Explorative scenarios are designed to allow for evaluation (1) of the expected future intensity of both hazards and (2) of their impact on the MRS (Sect. 6.4). The scenario analysis forms the basis for the development of measures in the field of urban land-use planning (Sect. 6.5) in response to the present exclusion in planning decisions of land-use and climate changes and their impact on hazard occurrence.

6.2 Methods and Data Sources

For the analysis of land-use changes and flood and heat hazard in the MRS, a series of raster and vector data is applied (Table 6.1). Sections 6.2.1 and 6.2.2 describe how the data is analysed in response to the research questions.

6.2.1 Analysis of Remote Sensing and GIS Data

As a first step, Landsat multispectral data (bands 1–5, 7) was classified using the *Erdas Expert Classifier*. Class descriptions are based in all cases on spectral, texture and content information from GIS (Geographic Information System) data (e.g., the river network). The same LULC classes were derived from each Landsat image from 2001, 2005 to 2009. Classes were generalized for further analysis as follows:

- Built-up areas (dense urban and peri-urban, intermediate urban and peri-urban, disperse urban and peri-urban)
- Green spaces (urban green spaces, sparse vegetation, woodland and grassland)
- Agricultural areas
- Barren land and open spaces
- Water (standing water and water courses)
- Snow

Since Landsat images do not cover the MRS entirely, its southernmost parts had to be omitted from the LULC analysis.

The next step consisted of a change detection analysis of the classification results, with prime reference to expansion of the built-up area, loss/gain of green spaces, and loss/gain of agricultural areas in the MRS. The generalized classes were recoded with numbers from 1 to 6, increasing by one order of magnitude at each point in time, e.g., agriculturally used areas were coded with 3 in 2001, with 30 in 2005, and with 300 in 2009. Adding up these values in each pixel produced an insight into the temporal development of each pixel with respect to land use.

Table 6.1 Overview of available data (*Source*: Authors' own presentation)

Data	Date (source)	Format/spatial resolution
Landsat 5 TM	December 7, 2001; February 1, 2005; February 12, 2009	Raster data, 30 m geometric resolution, bands 1–5 and 7 (thermal band 6: 120 m geometric resolution)
Municipalities	INE 2002	Polygon shape files, outlines of municipalities in the RM
River network	OTAS, 2005	Polygon shape files showing the main network of rivers and water bodies for the MRS, manually adapted for the four time steps using satellite data
Agricultural land	OTAS, 2005	Agriculturally used areas comprising crops and smaller pasture areas, manually adapted for each year using satellite data
Built-up areas	Own derivation based on Landsat 5 TM data	Built-up areas for 2001, 2005 and 2009 for the MRS including urban green and open spaces
Flood hazard zones	MINVU 1986, updated 2004	Polygon shape files showing outlines of areas with high (0–2 years return periods) flood hazards
	Perez (2009)	Raster data representing extent of flood hazard and flood depth for the San Ramón creek (La Reina)
Heat island map	Höfer (2011)	Polygon shape file at building block level containing areas with heat hazards on selected days in 2001 and 2009
Climatic predictions	Cortés et al. (2012)	Daily rainfall and temperature data, extrapolated for the period 2045–2065 using regional climate models and different IPCC scenarios
Modelled run-off values	Müller (2012)	Hourly run-off volumes modelled for the San Ramón creek in the MRS for different land-use scenarios and precipitation events

The spatial statistics of the land-use classifications for each municipality and in greater detail for flood and heat hazard zones were calculated in a GIS.

Thermal Landsat data (band 6) was analysed for two points in time (2001, 2009) to calculate areas affected by above-average temperatures. The grey values from the image were converted to brightness temperature values. These values were subsequently multiplied by an emissivity factor contingent on the land use of the pixel to obtain surface temperatures (Nichol 2009). In the course of this step the surface temperature map was resampled to the 30 m geometric resolution of the land-use map. Although this method does not deliver air temperatures identical to those measured by meteorological stations, which would indicate thermal stress more accurately, a satellite image produces a map with full spatial coverage of the MRS that could not be obtained by interpolating values from the scant weather-station network (compare Cortés et al. 2012). The temporal resolutions are undoubtedly one of its drawbacks. The image represents a snapshot taken at 10.30 a.m. local time and not the daily minima or maxima. It was not possible to track temperature curves throughout the day or comparisons between day and night. Nevertheless, it is the only reliable method of procuring an image for the entire study area. To derive a heat hazard map, a critical temperature threshold value was developed and applied. Statistics for built-up areas were interpreted. Defining the threshold value for above-average temperatures entailed specifying the sum of the mean temperature value over built-up areas and one standard deviation as the limit. The threshold value for the image in 2001 is 35 °C and 44 °C for the image in 2009, when overall surface temperatures were clearly higher in the whole of the study area. Next, the temperature value that occurs most often per building block was calculated and building blocks with temperature values above the threshold were masked out as hazard zones. The resulting hazard maps are used as a data base for further statistical calculations and analyses related to heat hazard.

6.2.2 The Application of Scenarios to Predict Future Land Use and Hazard Situations

Two explorative scenarios (BAU and CR, Chap. 3) depicting possible developments of flood and heat related conditions with the time horizon 2050 were designed for the purpose of estimating the impact of land-use and climate changes on hazard generation in the MRS (Sect. 6.4). During field work several interviews were held with local decision-makers in the area of land-use development and stakeholder workshops held to obtain expert knowledge of possible future urban growth. Factors relevant to land-use planning and development, such as population growth, the urban planning framework and the development of built-up areas and hazard zones, were described for the two future development alternatives (Sect. 6.4).

For the in-depth case study of the San Ramón basin located in the eastern part of the MRS in the Andean foothills, which is used as an example to quantify the impact of land-use and climate changes on flood hazards, two place-specific scenarios matching the framework scenarios were drawn up in map format. Estimated LULC changes in the basin were assigned manually using knowledge of the basin rather than, for example, a standardized reduction rate for vegetation.

6.3 The Relationship Between Land Use and Hazard Occurrence

This section outlines the current land-use pattern and previous developments (Sect. 6.3.1), focusing on hazard zones (Sect. 6.3.2) in the MRS, and establishes relationships to the occurrence of flood and heat hazard.

6.3.1 Land Use and Its Changes in the MRS Within the Previous Decade

Analysis of recent optical Landsat 5 TM data reveals that in 2009 the MRS contained predominantly dense and intermediate urban areas in the most central municipalities and medium-density areas on the urban fringe. The eastern part of the MRS is characterized by low-density residential areas with a higher proportion of green spaces. Green spaces around built-up areas currently function as buffer zones and natural flood protection. Agriculturally used areas with and occasionally without cultivation can be found on the northern, western and southern urban fringe, increasing towards the limits of the MRS (with the exception of the eastern limits).

The change detection analysis of land use in the MRS between 2001 and 2005, and between 2005 and 2009 (Fig. 6.1) indicates that the most significant changes in built-up areas in the first time span (2001–2005) occurred in Colina (+455 ha), Lo Barnechea (+399 ha), Lampa (+333 ha), Las Condes (+315 ha), and Buin (+263 ha). The most notable expansion between 2005 and 2009 took place in Lampa (+1,077 ha), Maipú (+608 ha), Puente Alto (+523 ha), Pudahuel (+462 ha), and Quilicura (+362 ha). Figure 6.1 distinguishes between strong changes (200 ha and above), moderate changes (between 50 and 200 ha), and no or light changes (50 ha and below). The groups refer to absolute changes per municipality rather than relative numbers. The figure highlights municipalities with moderate or strong expansion only.

The upsurge in built-up areas coincides with a loss of either agriculturally used land (Lampa, Colina, Maipú, Pudahuel, Calera de Tango, and Puente Alto) or green spaces (Lo Barnechea, Lampa, Puente Alto, Las Condes, Huechuraba). Loss of green space, however, is not necessarily a precursor of ongoing or planned construction activities. The impact of La Niña may also have led to generally dry conditions and vegetation loss. The year 2009 (represented by the latest satellite imagery used for this study), for example, showed rather dry conditions. On the whole, the number of green spaces in the entire MRS is, consequently, lower. It is therefore vital to take several LULC classes into account rather than to interpret one single change. The striking increase in green spaces in the region of the Andes in the second period (2005–2009) can be explained by the reduction in snow and glaciers. Areas previously covered in snow and now revealed to be barren land or to have sparse vegetation coverage belong in part to the green space category.

6.3.2 Current Land-Use Changes in Flood and Heat Hazard Zones

The proportion of green spaces in relation to built-up areas is low in most central municipalities of the MRS, which in terms of flood and heat hazard generation carries considerable risks. This section will explore the impact of land-use changes on both hazards as observed during the past decade, i.e., how urban expansion has affected the extent and location of hazard zones. The absence of data relevant to a quantitative analysis of flood hazard changes confined the analysis to qualitative and semi-quantitative aspects. In the case of heat hazard, multi-temporal data on hazard zones and land-use maps was used to establish a quantitative relationship between land-use changes and heat occurrence.

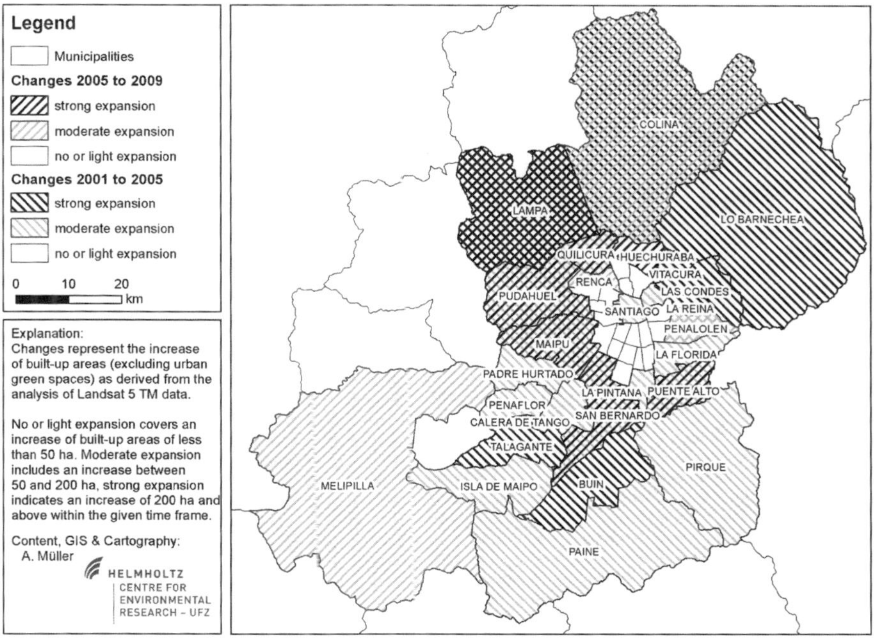

Fig. 6.1 Overview of municipalities in the MRS with moderate to strong urban expansion between 2001 and 2005, and between 2005 and 2009 (*Source*: Authors' own presentation)

The flood analysis is based on hazard maps officially available from the Ministry of Housing and Urbanism (MINVU, 1987, partly updated 2004), as no other data set was obtainable for the entire MRS. A first spatial analysis showed that in 12 of the 52 municipalities more than a respective 1,000 ha are highly flood prone, mostly in suburban municipalities.

Statistical analysis tools were then applied to provide information derived from multi-temporal Landsat 5 TM data on LULC in the highly flood-prone areas (return period 0–2 years) in the entire MRS in 2001 and 2009 (Table 6.2).

Numerous areas facing a high flood hazard are covered by urban built-up areas and agriculturally used land. A small percentage of these areas is covered by green spaces or barren land.

In a next step, a closer look was taken at the flood hazard situation in inner-urban areas, which although spatially less affected than the urban periphery are most likely to show a higher exposure of people and urban infrastructure to flooding. The municipalities of Quilicura, Lampa, San Joaquin, Maipu, and Peñalolén (compare Fig. 6.1) show the highest proportion of dense and intermediate built-up areas located in highly flood-prone zones (Table 6.3). Vitacura shows a high number of sparse urban areas in locations declared as highly flood prone in the PRMS.

The location of buildings in areas declared as flood prone does not automatically indicate a high exposure to the hazard, since exposure ultimately depends on the availability of protection measures. Updated information on current flood

Table 6.2 Land use and land cover in areas facing a high flood hazard in the MRS (in ha) (*Source*: Authors' own presentation)

LULC	2001	2009
Dense urban and peri-urban built-up area	1,970	1,833
Intermediate urban and peri-urban area	4,864	5,883
Disperse urban and peri-urban area	1,625	1,641
Urban and peri-urban green spaces	950	735
Green spaces outside urban areas	7,767	7,091
Agriculturally used areas	12,072	11,461
Barren land and open spaces	2,281	2,932
Water bodies	5,752	5,713
Snow	0	0
No data (undefined and background)	384	392

Table 6.3 Selected municipalities with a high number of built-up areas located in high flood hazard zones (2009) (*Source*: Authors' own presentation)

	Dense urban or peri-urban built-up area (ha)	Intermediate urban or peri-urban area (ha)	Disperse urban or peri-urban area (ha)
Quilicura	188	289	26
Lampa	181	457	62
San Joaquin	118	125	5
Maipu	117	386	41
Peñalolén	43	295	53
Vitacura	28	243	306

protection measures, however, is not included in the PRMS maps. Updating the flood hazard map and more precise access to flood hazard exposure would require additional field data. Although it counteracts planning regulations, construction in flood-prone areas is permitted once flood protection for the lot concerned is guaranteed, e.g., with structural measures (Carvacho 2010), in which case a more comprehensive and spatially broader impact analysis assessment is no longer mandatory. Consequently, areas known to be flood prone are becoming more densely populated. This leads first of all to a significant increase in exposure and the potential for damage of events exceeding the protection level. Secondly, it promotes the spatial extension of flood hazard zones, as less retention takes place and the flood hazard is shifted to locations in lower-lying regions. Hence although the impact of changing land-use patterns on flood hazard generation was not quantified, the adverse effects of urban expansion on areas prone to flooding or that have hitherto functioned as retention areas clearly points to past and current negative development. In sum, there is evidence of a prevailing increase in built-up areas in flood hazard zones with dramatic consequences for (1) hazard exposure and (2) the expansion of flood hazard zones with constant surface sealing (Table 6.2).

The LULC analysis was likewise carried out for areas exposed to heat hazard (2001 and 2009). Since the Landsat 5 TM optical bands were used for land-use delineation and the thermal band for surface temperature derivation, both land-use and surface temperature data was available for the same temporal and spatial coverage.

The results show that 481 ha of the areas affected in 2009 faced densification from medium to dense urban built-up areas and 330 ha densification from low to medium built-up areas between 2001 and 2009 (Table 6.4). An expanse of 422 ha in the newly affected areas showed a land-cover change from agricultural use with intermediate infructescence to one of low infructescence, i.e., loss of vegetation coverage. This proves that although the proportion of vegetation plays a central role in surface temperature development, it is not the sole determinant. Areas affected in 2001 and 2009 indicate changes in the density of urban and peri-urban built-up areas that involve both an increase and a decrease in vegetation (compare Table 6.4). However, the substantial amount of space that was converted from dense to medium urban built-up areas between 2001 and 2009 and experienced a reduction in surface temperatures between both points in time underlines a clear causal relation between vegetation coverage and affectedness (Bowler et al. 2010; Gill et al. 2007).

Figure 6.2 displays the areas exposed to heat hazard and the land-use map from 2001. Geostatistical analysis shows that in most municipalities the land-use types most exposed are dense and intermediate built-up areas (Table 6.4). In peri-urban municipalities such as Lampa and Colina, the built-up areas most exposed are densely populated, while only a small percentage of the intermediate and sparse category suffers exposure. In the eastern part of the MRS exposure is minimal as these areas are rarely affected by heat hazards.

While temperature values represent the temperature that occurs most often for the entire building block with an average size of 100 × 100 m (cf. Sect. 6.2), the land-use information refers to single image pixels sized 30 × 30 m. This should be taken into consideration when analysing more diverse land-use information. Consequently, only the three main exposed land-use classes were derived for each municipality. Among those are dense, intermediate, and sparse urban and peri-urban built-up areas with a clear affectedness dominance of dense urban built-up areas. This is a logical consequence of heat hazard acceleration in densely built-up areas. A small number of urban green spaces, areas with sparse vegetation, barren land, and cultivated land with little or no vegetation coverage also show above-average surface temperatures, since the cooling effects of vegetation depend on plant physiology. Grassland, peri-urban green spaces, woodland and water are therefore not affected by heat hazard occurrence or to a minor degree only.

Figure 6.2 shows three macro views of exemplary areas. The first zoom-in indicates that dense peri-urban built-up areas and uncultivated land in the northern suburbs (Colina) are more affected by heat hazard than built-up areas with intermediate density (high temperatures marked by black hatched polygons). For methodological purposes, however, the analysis includes inhabited areas only, i.e., agriculturally used areas in the surroundings of the settlement in zoom-in 1 are excluded. Likewise, the second zoom-in shows that areas with a higher building

Table 6.4 Land-use changes in areas with above-average temperatures (*Source*: Authors' own presentation)

Land-use changes in areas affected in 2009 only	
Intermediate to dense urban built-up area	481 ha
Agriculturally used areas: intermediate to low infructescence	422 ha
Sparse to intermediate urban built-up area	330 ha
Land-use changes in areas affected in 2001 only	
Dense to intermediate urban built-up area	1254 ha
Dense to intermediate peri-urban built-up area	210 ha
Agriculturally used areas: low to intermediate infructescence	193 ha
Land-use changes in areas affected in 2001 and 2009	
Dense to intermediate urban built-up area	561 ha
Intermediate to dense urban built-up area	342 ha
Dense to intermediate peri-urban built-up area	57 ha

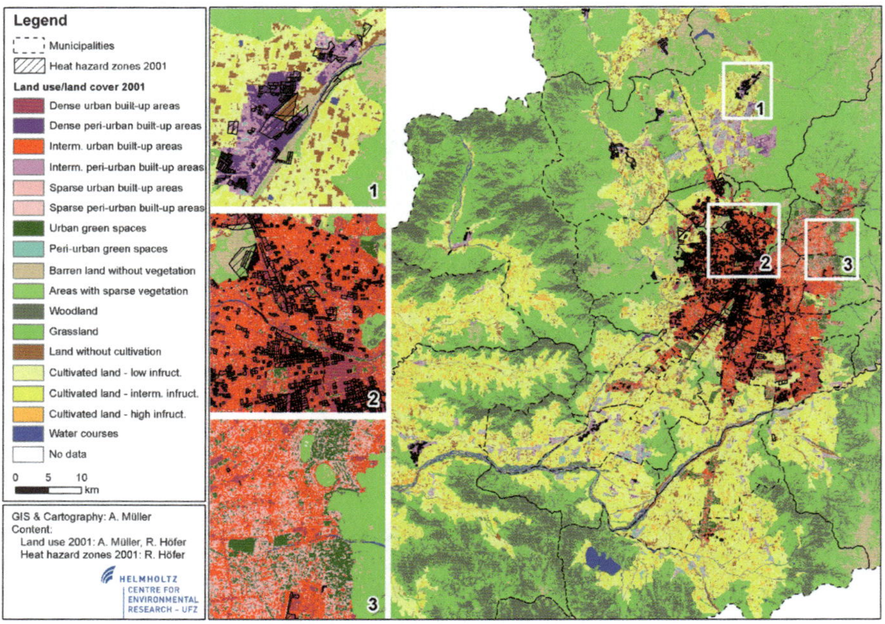

Fig. 6.2 Location of areas exposed to heat hazard in 2001 and an LULC map from 2001 (MRS) (*Source*: Authors' own presentation)

density (coloured dark red) are affected more often than intermediate density or non-built-up areas. This is not a one-to-one relationship. As shown for the most central areas of the city (lower left region of zoom-in 2), high density built-up areas suffer only partially from above-average temperatures owing to shade provided by surrounding high-rise buildings. Other components such as orography and wind also influence temperature, albeit they are excluded from this analysis. The eastern part of the city along the Andean piedmont (zoom-in 3) shows an overall more favourable situation. Dense built-up areas in this part of the city are sparse and temperatures lower.

Figure 6.3 illustrates changes in the location of heat hazard between 2001 and 2009, again at building block level. Although the derived surface temperatures vary significantly between both acquisition dates in 2001 and 2009 (threshold value 2001: 35 °C, 2009: 44 °C), the distribution of heat hazard in the MRS is for the most part consistent, i.e., the pattern of affected areas remains stable along the broad axis from Quilicura in the north across the city centre to San Bernardo and Puente Alto in the south. Densely built-up municipalities located to the west of this axis are likewise affected. The areas newly threatened by heat hazard in 2009 are primarily located in Pudahuel, San Bernardo, and Puente Alto. In contrast, Quilicura, Cerrillos and the central municipality of Santiago managed to reduce heat hazards, at least as shown for the snapshot moment of the satellite overflow. This positive effect is most likely the result of an increase in green spaces in the respective areas, for example the land-use changes in Cerrillos associated with the closure of the airport and its interim use as urban brownfields. The eastern municipalities along the piedmont are rarely affected by heat hazard due to a higher proportion of green spaces. The sole exception to this pattern is the densely populated municipality of Puente Alto, which belongs to the areas with a lower social status and a major growth in built-up areas. In 2009 it was heavily affected.

6.4 Development of Future Land-Use Patterns

The following contextual storylines anticipate possible alternative development of land use and climatic conditions, and of selected MRS regulations relevant to the development of heat and flood hazards.

The **Business As Usual** (**BAU**) scenario makes the following assumptions:

- According to the National Statistics Institute, the municipalities with the highest increase in **population** between 1990 and 2020 (expected) are Quilicura (+*622* %), Maipú (+*393* %), Puente Alto (+*270* %), Lampa (+*253* %), Calera de Tango (+*191* %), Lo Barnechea (+*177* %), Colina (+*167* %), and Pirque (+*156* %) (INE 2011). Consequently, **urban growth** continues to expand in the direction of the rural areas and the eastern municipalities located along the Andean foothills.
- Growth towards the eastern periphery continues only slightly after 2030, when the current **construction limit** of 1,000 masl is lifted. The respective municipalities remain stable to a large extent until 2050, with gated communities developing in urban areas over 1,000 masl after 2030. The urban expansion process towards the north and south can be characterized as leap-frog development, where outlying areas are taken up with residential, commercial and industrial development (e.g., Lampa, Colina, Melipilla, Pirque, San Bernardo, Buín).

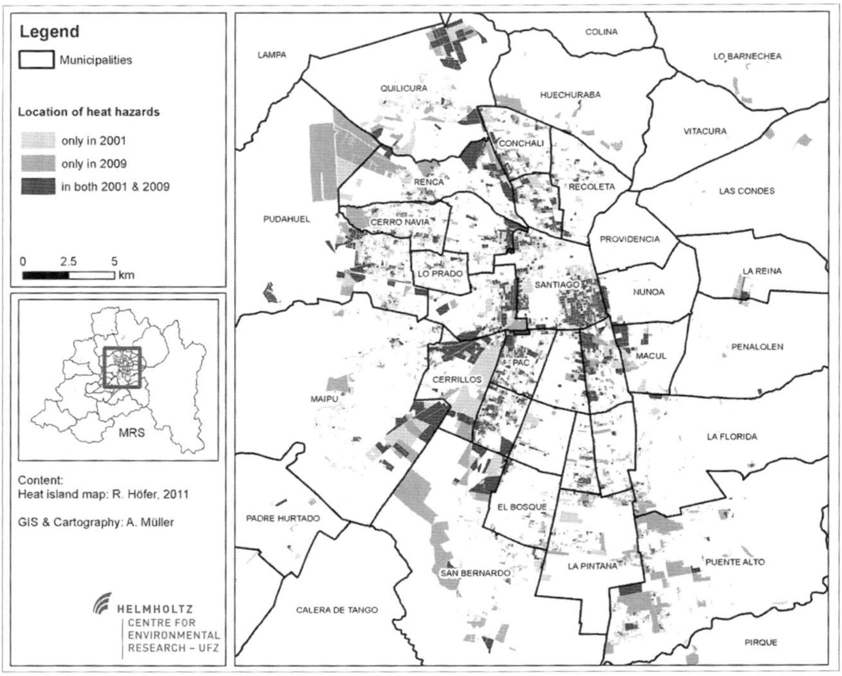

Fig. 6.3 Location of heat hazards in 2001, 2009, and for both years in the central MRS (*Source*: Authors' own presentation)

- The remaining **agriculturally used land** is partly converted into residential areas to accommodate new dwellers, primarily along the southern, northern and western limits of the city.
- Urban expansion leads to a vast increase in impervious surfaces, while private **green spaces** are slow to develop due to small lot sizes and high land prices. Public green spaces are maintained although water for irrigation has become a scarce resource and more expensive. In terms of climate development according to the B2 scenario, lower stream flow data is the dramatic outcome of urban expansion (Cortés et al. 2012). A shift in the precipitation pattern leads to a negative water balance in the Maipo basin in the summer, equalling out in the winter and spring in the course of processes of precipitation and snow melt (Chap. 4). Since native species remain highly limited in the urban areas of the MRS, the level of irrigation activities for urban green spaces is on the increase. Consequently, neither the amount nor the quality, i.e., the vitality, of green space is increasing.
- **Urban planning** is poorly coordinated. There is a prevailing disregard for ecosystem functions such as retention areas and ventilation corridors, and an absence of appropriate risk assessments. Although the creation of new green spaces is envisaged up to 2030 in the renewal of the PRMS (2008), the objective

is to reach the international standard of 9 m^2 per inhabitant rather than to minimize hazards. Changing climatic conditions are not taken into account in the urban planning process, i.e., the plan is not updated with respect to projected climate change.

The **Collective Responsibility** (**CR**) scenario describes an alternative future following the framework scenarios presented in Chap. 3 and is based on the following assumptions:

– The birth rate in the MRS is in decline and the urban **population** growth rate stagnant. There is little in-migration from outside. On the contrary, a percentage of the urban population of Santiago de Chile moves to other cities in Chile as a result of a variety of decentralization efforts on the part of the government. The development of medium-sized towns within the MRS is heavily regulated to prevent smaller towns from accreting to a large urban body. Inner-urban migration within the MRS, on the other hand, leads to the desired process of urban densification, as living on the fringes is no longer an attractive option (Hölzl et al. 2011).
– The urbanization process is characterized by a slowing down of **urban growth** into the rural areas as a result of strong participation of civil society and NGOs in urban planning decision-making. The negative impact of extensive urban expansion on environmental functioning in the MRS, especially in times of climate change, has fostered re-urbanization of urban core areas in the AMS and the current **construction limit** of 1,000 masl is maintained. Multi-storey buildings dominate the central areas of the city. Existing houses are being restored and renovated. To improve the compactness of the city, which is the key aim of settlement development, gaps between houses are closed. The municipalities with vertical growth include Quilicura, Maipú, Lo Barnechea, Huechuraba, and Puente Alto. Hence the urban body becomes far more compact with shorter distances to cover.
– The **agriculturally used areas** along the southern, western and northern urban fringe, as well as the natural areas that form a buffer around the built-up body are to a large extent maintained.
– Urban **green spaces** are designed in such a way as to include a higher proportion of native vegetation and thus reduce irrigation efforts. The maintenance of public green spaces and the creation of green corridors gains in significance and attention. Investment in public green spaces is a major development goal of the state and includes investment in technologies that make use of grey water for irrigation purposes. As a method of easing the protection of ecologically valuable areas, the state pays compensation to the respective municipalities.
– Although the multitude of actors involved in **urban planning** processes remains, coordination and synchronization has improved. Functions and responsibilities are more clearly identified. The legal tools available to the planning process are availed of in a coordinated manner and lead to satisfactory communication and exchange of information among the actors concerned. A regional development strategy is established and based on joint decisions by the Regional Government (GORE), the regional offices of the Ministry of Environment (SEREMI-Medio

Ambiente), the Ministry of Housing and Urbanism (MINVU), private actors and civil society. The result is a moderate and concerted urban development that takes changing environmental conditions into account.

6.4.1 The Impact of Future Climate and Land-Use Change on Hazard Occurrence in the MRS

The scenario analyses in this section combine the conditions of the projected climate change (cf. Chap. 2) and land-use developments presented in Sect. 6.3 to deliver assumptions about future flood and heat hazard development in the MRS. Quantitative assumptions are made for the San Ramón case study catchment.

The total amount of precipitation will decrease by 20–100 mm annually in the future. Only the intensity of extreme events will show a slight increase for the B1 scenario (events with 20 mm and above) (cf. Chap. 2, Cortés et al. 2012). In other words, flood probability is growing for low frequency events but remains stable or is in decline for frequent events. An overall reduction in rainfall intensities combined with higher temperatures and drought periods, however, has an adverse effect on the amount and vitality of natural green space. In addition, extended surface sealing predicted for the BAU scenario will lead to less soil infiltration capacities and a considerable increase in surface run-off and floods. Thus under BAU scenario conditions areas affected by floods will increase by 5–10 % up to 2050 along the piedmont, the location of smaller watersheds that today frequently lead to riverine floods (Müller 2012; Perez 2009).

To gain a more quantitative insight into the behaviour of these small watersheds, an in-depth study for the San Ramón Creek located in the municipalities of Las Condes and La Reina was carried out.

Applying the hydrological precipitation-run-off model HEC-HMS to the catchment allowed for quantification of expected land-use changes with regard to watershed run-off volume and consequently to flood hazard in the adjacent urban area. Two spatially explorative LULC scenario alternatives (scenarios I and II) that fit the scenario assumptions of BAU and CR were developed. Scenario I (BAU) is based on loss of vegetation due to drier summers and higher average temperatures (–20 % sparse vegetation = –1.47 km^2, –44 % woodland = –0.44 km^2, +19 % = –1.9 km^2 barren land); scenario II (CR) assumes afforestation activities (+153 % = +153 km^2 at the expense of areas with sparse vegetation) along the urban fringe to increase vegetation (see Müller 2012 for more details on the modelling process).

It is evident from the model outcome that the land-use scenario poses a threat, i.e., loss of vegetation due to longer periods of drought combined with heavier rainfall during extreme events (BAU). Afforestation, a possible development in the CR scenario and a preferable alternative, should, however, comprise native species that adapt to low water availability during the dry season.

In a next step, changing run-off levels were related to the simulated flood hazard maps to investigate how changes in the run-off affect flood hazard. Six flood hazard maps specifically derived for the San Ramón channel were available from a previous study (Perez 2009). For an event with a 5-year return period, the affected urban area would expand from 188 to 214 ha (run-off +7.8 m^3/s) (Perez 2009). With an increase from 50.9 to 64.6 m^3/s (approx. a 30-year return period), flooded areas would grow from 243 to 396 ha (Perez 2009). Although the changes are non-linear, they indicate that damage to the dense urban environment would magnify. In other words, absolute changes are higher for high run-off values, i.e., low frequency events. Although these figures seem harmless at first sight, they are high when referred back to the research area, i.e., an urban environment with a high density of people and values. In addition to a widening of the affected areas, most flooded areas would face a higher water depth. This in turn would require further investment in measures to reduce physical exposure and most likely result in longer flood periods. Hence hazard occurrence will reflect socio-economic differences more clearly. Other regions of Santiago de Chile outside the case study area, particularly along the Andean foothills, are affected by riverine floods from creeks similar to San Ramón or by urban floods. The flood affected areas remain constant along the main Mapocho and Maipo Rivers, since they are subject to high priority mitigation. In the CR scenario, ample funds are invested in the maintenance and extension of storm water infrastructure, leading to a decline in urban floods throughout major parts of the city.

With respect to the development of heat in the MRS, it is expected that the warmer summer temperatures of +1.5 to 2 °C in both scenarios—with slightly higher temperatures in scenario A1—will exacerbate the hazard further, especially after 2030 (Cortés et al. 2012). For the Santiago station located in the central part of the MRS, the number of days above 30 °C will increase from approximately 55 days today to 76 days up to 2030 and 88 days up to 2050. The projections are lower for scenario B1 with approximately 69 days up to 2030 and 76 days up to 2050, but heat waves will generally last for a higher number of days, especially along the affected area from Quilicura to San Bernardo and Maipú (Cortés et al. 2012). There will be an increase on the periphery of between 20 and 40 days.

The findings in Sect. 6.3.2 clearly underpin the strong relationship between loss of green spaces and the increase in sealed surfaces and surface temperatures in the MRS. This development will continue, not least in areas with a predicted population growth. Hence areas affected by heat hazard under BAU scenario conditions are estimated to expand by approximately 10 %, with the primary increase in densely populated areas that accommodate households from the lower social strata. Heat will therefore pose a threat in the summertime to a growing number of people who are both exposed and vulnerable. Higher temperatures also accelerate irrigation needs in the urban area and will lead to a reduction in natural green spaces as observed in the past in drier years.

The situation is slightly more relaxed under CR scenario conditions. Here a higher percentage of native species and the use of grey water for irrigation purposes facilitates green space maintenance. While a dense urban environment, which goes hand in hand with the compact city model, normally fosters urban heat hazards, installation measures such as green walls and maintenance of ventilation corridors reduce these adverse effects. Built-up populated areas affected by above-average temperatures will increase by merely five per as a result of effective mitigation measures. Only low-income areas in the western and southern parts of the built-up body will experience more negative effects. Furthermore, the conversion of a number of public lands (e.g., railroads, train stations, airports and disused military installations) to green infrastructure enables air temperature control via improved ventilation patterns and increased evapotranspiration.

6.5 Adaptation Measures

The analyses in Sect. 6.4 point to the need for prompt action when it comes to adapting to climate change and changing hazard conditions in the MRS. This section will propose and discuss three specific measures to reduce the flood and heat hazard potential in the course of urban growth and climate change in the MRS. The development of these measures was sustained by an intensive scoping study conducted by a Chilean expert.

The measures were selected for their effectiveness but also had to fit in with local conditions. They are designed as medium- to long-term strategies but located at different levels. The first measure, a web-based information data base, addresses a national ministry and its regional department; the second measure, the introduction of a green factor for new developments, is a planning policy at regional or municipal level. The third measure, the re-activation of irrigation channels, is a place-specific and localized measure valid for specific areas of the MRS only, and addresses ministries, NGOs and local residents.

6.5.1 Web-Based Information Data Base

One of the principal constraints on the development of measures for adaptation to climate change is the poor availability of and access to relevant information and existing data sets in the MRS. A useful solution would be to **establish a web-based data base with a map component that compiles and provides existing information and allows for the generation of new information via query functions or overlay analysis**. It should focus on natural units such as river catchments and involve ministries responsible for various types of infrastructure and urban planning, and their local departments. The information must be accessible to the decision-makers, planners, and/or engineers concerned. This measure was selected because it

1. Overcomes the fundamental problem of limited data availability
2. Would support the dissemination and application of knowledge collected in local case studies
3. Would foster collaboration between different institutions at horizontal and vertical level

4. Would support and encourage the development of further measures and the inclusion of environmental information in decision-making processes

Ideally, the system builds on the three existing map services in the relevant ministries. The installation of such a system requires no administrative changes and is coherent with existing laws, including the National Action Plan on Climate Change (PNACC). This means it could be put into practice within a short period of time.

In terms of barriers, it is problematic that most of the data currently available is collected by ministries at national level and on a small scale only. It furthermore represents a sectoral perspective and is not always collected for the purpose of producing information on the current status of the ecosystem or conducting complex environmental analysis. The scale and year of origin tends to be incoherent and the issue of copyright needs to be clarified.

6.5.2 Green Factor for New Public and Commercial Developments

The importance of green spaces for heat and flood hazard reduction was pointed out in Sect. 6.3. To define a minimum amount of green spaces for each municipality or indeed each census district, the introduction of a **green factor for new public and commercial developments** is proposed. The measure was selected because it

1. Would improve storm water management
2. Presents an opportunity to compensate surface sealing with the creation of new or maintenance of existing vegetation where new settlements evolve
3. Has been shown to be an effective measure internationally (City of Malmö 2011; City of Seattle 2011; Senate Department for Urban Development and the Environment 2011; Kazmierczak and Carter 2010)
4. Could contribute in the case of Santiago to the improvement of air quality

The norm applies to construction that exceeds 400 ha or comprises more than 20 parking lots. The validity and effectiveness of the project is defined by a minimal relation between sealed surface and green area. The type of green space is not specified but offers a range of options from green roofs through planting new trees to preservation of existing grassland.

The measure would require contributions from various ministries and an amendment to the General Law of Urbanism and Construction. It might also interfere with existing municipal development plans—a key constraint. Modifications of existing laws are long-term processes in the study area. Thus, the voluntary introduction of a green factor to new construction should be accompanied by a tax advantage. Funding the measure remains a crucial factor, as suitable incentives need to be found. The maintenance of green spaces amounts to approximately US$6 per m^2 per month, a costly operation for a municipality or a private investor (Carvacho 2010). Municipalities that accommodate people from the lower socio-economic strata and dispose of a low budget can rarely afford the necessary maintenance. Ideally, the affected public would develop ecological sensitivity and a willingness to participate in a green adaptation process of this kind.

6.5.3 Re-Use of Existing Irrigation Channels

The third measure is site-specific and touches on one of the causes of flood generation. Flooding along the Andean piedmont frequently occurs as a result of the low capacity of the waterways. In the past, the area along the foothills was used for agricultural purposes and had an efficient irrigation system consisting of a channel network. Although urban expansion saw a decline in the number of fields used for agricultural purposes, several channels still exist and can be utilized for storm water evacuation in the case of strong precipitation events. This would lower the amount of run-off in the creeks along the piedmont and reduce flooding. Hence **reintroducing the existing irrigation channel system along the Andean piedmont** is a medium-term structural measure that targets flood hazard specifically and minimizes its consequences. The measure was proposed as a result of field observations and comprises an inventory of the existing irrigation channels system and its completion where necessary. The main obstacle to its implementation is the stipulation in the water regulations that irrigation channels should not be used to collect or dispose of storm water. The solution calls for an agreement that establishes the proposed measure as an exception to the current regulations, to be signed by the private owners of the channels, the NGO *Canalistas de Maipo*, and the relevant institutions.

6.6 Conclusions

This chapter analysed the impacts of climate and land-use changes on flood and heat hazard in the MRS. It first explained the relationship between land-use developments in the context of urban growth, climate change, and the development of heat and flood hazards. It then explored land-use changes in the MRS in the last decade to determine their site-specific influence on changes in flood and heat in the study area. It was shown that both hazards are negatively affected by the ongoing process of urban expansion, surface sealing, and the decline in the amount and quality of green space.

Subsequent to highlighting the interlinkages between the previous decade and the current state, explorative scenarios were developed to determine the future impact of land-use and climate changes on both hazards. The findings show that although the impact of expected land-use changes is more dramatic in the BAU than the CR scenario, existing hazards will not be minimized in any of the scenarios. The high consumption of land for new built-up areas in the BAU scenario reduces the vegetated area significantly. As a result, both flood hazard and the occurrence of heat islands will grow. In the CR scenario, afforestation activities reduce the amount of run-off that reaches the city. They can balance out vegetation loss resulting from longer drought periods in the catchment areas up to a point but cannot compensate for deficits in the MRS as a whole. Only if construction restrictions are upheld will the urban periphery maintain its function as a buffer zone and minimize river flood hazards.

Population growth and risk generation go hand in hand with the location of a growing number of people, buildings and infrastructure in hazard-prone areas, occasioning a permanent increase in the exposure of values to flood and heat. Although new building sites do not necessarily evolve in hazard-prone areas, they can intensify existing hazards and lead to a higher exposure of neighbouring lots. This notwithstanding, current construction activities take place in areas declared in the PRMS as hazard prone, since flood-prone areas are economically compelling and construction in these areas is permitted by law following compliance with specific conditions (Carvacho 2010).

Ongoing and expected future developments clearly call for immediate action by local stakeholders. Concrete measures for adaptation should be designed and put into practice. Three measures were proposed for the reduction of flood and heat hazard. Although they have the potential to reduce the root causes, these measures are bound up with institutional barriers that need to be overcome.

References

Barton, J., Kopfmüller, J., & Stelzer, V. (2011). *Quantitative variables: Tendencies and scenario estimations 2010–2050*. Santiago de Chile Karlsruhe: Internal Report.

Bowler, D. E., Buyung-Ali, L., Knight, T. M., & Pullin, A. S. (2010). Urban greening to cool towns and cities: A systematic review of the empirical evidence. *Landscape Urban Plan, 97*, 147–155.

Carvacho, D. (2010). Recuperación de áreas inundables urbanas. Caso: Cuidad de San Vicente de Tagua Tagua. Master's thesis, Universidad de Chile.

City of Malmö (2011). The green city. Stormwater and green roofs, http://www.malmö.se/English/Sustainable-City-Development/Augustenborg-Eco-City/The-Green-City.html. Accessed 21 December 2011.

City of Seattle (2011). Seattle green factor. What is the Seattle green factor? Department of Planning and Development, http://www.seattle.gov/dpd/permits/greenfactor/Overview/. Accessed 23 December 2011.

Cortés, G., Schaller, S., Rojas, M., Garcia, L., Descalzi, A., Vargas, L., McPhee, J. (2012). Assessment of the current climate and expected climate changes in the Metropolitan Region of Santiago de Chile. UFZ-Report 03/2012, Helmholtz Center for Environmental Research – UFZ, Leipzig.

Direccíon General de Aguas (2009). Servicios satelitales en tiempo real. http://dgasatel.moptt.cl. Accessed 15 March 2009.

Gill, S., Handley, J., Ennos, A., & Pauleit, S. (2007). Adapting Cities for climate change: The role of the green infrastructure. *Built Environment, 33*, 115–133.

Höfer, R. (2011). The derivation of heat islands from remote sensing data in Santiago de Chile. Internal Working paper, unpublished.

Hölzl, C., Krellenberg, K., Heinrichs, D., Welz, J., & Kabisch, S. (2011). *How sustainable are processes of social and spatial differentiation in Santiago de Chile? Current situation and future scenarios for social inclusion, UFZ Report 02/2011*. Leipzig: Helmholtz Center for Environmental Research – UFZ.

Instituto Nacional de Estadísticas (2002). Censo Nacional 2002. Santiago de Chile.

Instituto Nacional de Estadísticas Chile. Proyecciones de población al 30 junio. 1990 – 2020. Región Metropolitana de Santiago: Comunas, sexo y edad. http://www.ine.cl/canales/chile_estadistico/demografia_y_vitales/proyecciones/DatCom/SalComUsuarios-13Tok.xls#'!A195. Accessed 5 July 2011.

Kazmierczak, A., & Carter, J. (2010). *Adaptation to climate change using green and blue infrastructure. A database of case studies*. Manchester: University of Manchester.
MINVU (2007) Evolution of urban growth. Vector data set.
Müller, A. (2012). Areas at Risk - Concept and Methods for Urban Flood Risk Assessment. A case study of Santiago de Chile. Megacities and Global Change, Kraas, F., Heintzenberg, J., Herrle, P., Kreibich, V. (Eds.), No. 3, Franz Steiner Verlag, Stuttgart.
Müller, A., Reiter, J., & Weiland, U. (2011). Assessment of urban vulnerability towards floods using an indicator-based approach – a case study for Santiago de Chile. *Natural Hazards and Earth System Sciences, 11*, 2107–2123.
Nichol, J. (2009). An emissivity modulation method for spatial enhancement of thermal satellite images in urban heat island analysis. *Photogrammetric Engineering and Remote Sensing, 75*, 547–556.
Perez, J. (2009). Riesgo de inundación producto del cambio climático, Caso de estudio: Quebrada de Ramón, Master's thesis, Universidad de Chile.
Plataforma Urbana (2011). Finalmente se aprueba el PRMS 100. http://www.plataformaurbana.cl/archive/2011/03/31/finalmente-se-aprueba-el-prms-100/. Accessed 31 March 2011.
Romero, H., Ihl, M., Rivera, A., Zalazar, P., & Azocar, P. (1999). Rapid urban growth, land-use changes and air pollution in Santiago, Chile. *Atmospheric Environment, 33*, 4039–4047.
Romero, H., Vásquez, A. (2005). Evaluación ambiental del proceso de urbanización de las cuencas del piedemonte andino de Santiago de Chile. EURE, XXXI:97-118.
Senate Department for Urban Development and the Environment (2011). A green city center – BAF – Biotope area factor, http://www.stadtentwicklung.berlin.de/umwelt/landschaftsplanung/bff/index_en.shtml, Accessed 21 December 2011.

Understanding Hazard Exposure for Adaptation in a Climate Change Context

7

Juliane Welz, Anke Schwarz, and Kerstin Krellenberg

Abstract

The literature reveals that marginalized groups are more exposed to hazards at their place of residence than other groups. Given the patterns of profound social inequality in Santiago de Chile and ongoing processes of socio-spatial differentiation, it could be assumed that the residents most exposed to hazards associated with climate change belong to the lower socio-economic strata. The research analysis of city-dweller exposure to flood and heat hazard, using innovative distributional indices, provides empirical evidence that in the case of Santiago residents from all social strata are exposed in one way or another. The present study shows the overall hazard exposure for the Metropolitan Region of Santiago de Chile (MRS), highlights the population groups most exposed to hazards and depicts inequalities in residential patterns with respect to socio-economic status and physical housing conditions. Finally, adaptive measures customized to suit existing legal and institutional frameworks are proposed and discussed in the pursuit of hazard exposure reduction.

Keywords

Hazard exposure • Distributional patterns • Adaptation measures

7.1 Introduction

It is widely stated in the literature that marginalized groups are more exposed to hazards at their place of residence than others (cf. Satterthwaite et al. 2007; Wisner et al. 2004). Against the backdrop of persistent patterns of social inequality (CEPAL 2011; UN-Habitat 2012) and limited access of the lower socio-economic

J. Welz (✉) • A. Schwarz • K. Krellenberg
Department of Urban and Environmental Sociology, Helmholtz Centre for Environmental Research – UFZ, Permoser Str. 15, 04318 Leipzig, Germany
e-mail: juliane.welz@ufz.de

K. Krellenberg and B. Hansjürgens (eds.), *Climate Adaptation Santiago*,
DOI 10.1007/978-3-642-39103-3_7, © Springer-Verlag Berlin Heidelberg 2014

classes to housing and land markets, the question arises whether this observation applies specifically to the Latin American city. Empirical research on the socio-economic characteristics of populations exposed to climate change related hazards, however, has been quite limited for the Latin American urban context (cf. Leite da Silva et al. 2009).

This chapter focuses on the implications of hazards related to climate change in Santiago de Chile, where processes of socio-spatial differentiation are apparent (Hölzl et al. 2011; Rodriguez and Arriagada 2004; Rodriguez 2001; Sabatini et al. 2001) and housing allocation by household socio-economic status heavily influenced by market forces (Sabatini and Salcedo 2007; Hidalgo 2004). Does this immediately imply that Santiago's poor are among those most at risk from climate change related hazards? The research presented here is the result of an analysis of city-dweller exposure to flood and heat hazards, and provides empirical evidence of striking socio-economic differences in hazard exposure for Santiago de Chile, which cannot simply be reduced to higher exposure of 'the poor'.

Taking into account both residents' socio-economic status and their physical housing conditions, the chapter analyses the impacts of climate change related hazards such as flood and heat on the population of the Metropolitan Region of Santiago de Chile (MRS).

A contextual vulnerability approach is adopted, interpreting vulnerability to climate change effects as a human rather than a physical condition, caused by the impact of hazards and people's specific social and economic circumstances. When vulnerability is understood as a social product embedded in a particular social, economic and political setting (Wisner et al. 2004), sensibility for the context becomes a starting point for adaptation strategy development. Looking at the MRS, the interplay of urban municipalities and those with a more rural hinterland reflects the complexity and multiple interconnections and interdependencies of actors, assets and infrastructures that frame vulnerability. Although it focuses exclusively on an exposure analysis, the urban vulnerability concept of Kuhlicke et al. (2012), which explicitly addresses this complexity, is adopted.

Following the hypothesis that exposure to flood and heat hazards in the MRS (cf. Chap. 6) differs according to residents' socio-economic status and physical housing conditions, this chapter contrasts urban and rural municipalities in terms of predicted climate change impacts. The analysis brings the likelihood of exposure to flood and heat to the fore, following a set of research questions:

- What is the distribution pattern of exposure in the MRS according to residents' socio-economic status and physical housing conditions?
- What tendencies can be identified in the light of expected future distribution patterns of exposure to hazardous events?
- What are the context-specific recommendations for adaptation to flood and heat hazard under the predicted climate change?

The chapter is organized as follows: Section 7.2 clarifies the concept of 'urban vulnerability' within the regional context of MRS and advocates adaptation action. Section 7.3 explains the methodological framework and operationalization scheme, and examines a set of applied indicators. Sections 7.4 and 7.5 illustrate the status quo assessment of urban and rural areas affected by flood and heat hazard, focusing on the exposed population's socio-economic situation and physical housing conditions. Section 7.6 estimates future trends in hazard exposure, creating climate and urban development scenarios for the year 2050. Section 7.7 outlines adaptation strategies. A concluding Sect. 7.8 summarizes the main findings on hazard exposure and the potential for hazard reduction in the MRS.

7.2 Understanding Urban Vulnerability and the Need for Adaptation

Vulnerability can essentially be conceptualized as the outcome of environmental, social, cultural, institutional and economic structures and processes, leading to the unequal distribution of capacities and resources (Chambers 1989; Adger and Kelly 1999). The concept of 'urban vulnerability' goes a step further and highlights urban complexities, bearing in mind the numerous interdependencies between the diverse spheres of urban life and the environment (Kuhlicke et al. 2012). By adding the notion of 'urban', Kuhlicke et al. (2012) provide a vulnerability definition referring to the degree to which individuals, infrastructures and physical assets in urban environments are exposed or susceptible to environmental hazards, as well as to their capacity to cope with and adapt to the negative impacts. This definition is based on the general concept of vulnerability as it emerged in the hazard and disaster community: vulnerability is understood as "the characteristics of a person or group and their situation that influence their capacity to anticipate, cope with, resist and recover from the impact of a natural hazard" (Wisner et al. 2004:11). It takes three dimensions into account: exposure, susceptibility, and coping and adaptive capacity:

- *Exposure* refers to the physical precondition to be affected by natural hazards, thus acting as a bridge between the natural and social science approaches (Fuchs et al. 2011). The present study analyses the exposure of Santiago's population to flood and heat hazard.
- *Susceptibility* denotes the precondition to suffer harm due to a certain level of fragility or other disadvantageous conditions. A household with a weak social network, for instance, may lack the appropriate warning system for a hazard occurrence (Kuhlicke et al. 2011).
- *Coping and adaptive capacity* describes the ability to prepare for, cope with and recover from a hazard impact. This capacity is characterized by the presence (or absence) of a set of assets or resources (political, physical, natural, social, financial) of, for example, an individual or group (Chambers 1989). Coping capacity is thus directly linked to people's access to crucial resources such as employment, education, disaster risk management programmes or integration into strong social networks.

The concept of vulnerability is broadly used in the domain of climate, social and disaster research to quantify the impact of climate change and the assessment of its attendant risks. According to Klein et al. (2007), dealing with adaptation requires an understanding of the vulnerability of societies and ecosystems to the impacts of climate change, of their capacity to respond, and of the socio-economic cost of adapting to them. Risk reduction and the generation of 'adaptive capacity' in an effort to decrease overall vulnerability to potential climate change impacts is the general purpose of adaptation to climate change as defined by the IPCC (Adger et al. 2007). A vulnerability assessment is therefore essential to determine the groups most vulnerable to such impacts and to pursue the reasons why (Ribot et al. 2009).

This chapter focuses on assessing vulnerability to climate change effects as an add-on to existing vulnerabilities that are structural in nature and frequently linked to poverty and inequality. In the case of the MRS, priority is given to hazard exposure. In our understanding, hazard exposure is not driven by physical changes to climate alone, but also by natural and anthropogenic factors such as demographics, economics, landscape and infrastructure, all of which impact on the sensitivity of places and populations to climatic change and their capacity to respond. Hence the present study will indicate the overall hazard exposure in the MRS, call attention to the population most exposed to hazard and depict inequalities in the patterns of hazard exposure with reference to residents' socio-economic status and physical housing conditions. Although there is an apparent lack of research on the susceptibilities to climate change induced hazards in Santiago de Chile and the relevant coping capacities, it would exceed the scope of this chapter to attempt to fill the gap on these components of a comprehensive vulnerability analysis. The coping and/or adaptive capacities highlighted here refer instead to institutional responses to climate change. Adaptive measures for hazard exposure reduction, custom-built for implementation in existing legal and institutional frameworks, are proposed and discussed.

Against this backdrop, the urban vulnerability analysis for the MRS concentrates on the following aspects:

- Flood and heat having been identified as two hydro-meteorological hazards that impact on the MRS and are aggravated by ongoing urbanization, climate change and, not least, their interlinkages (cf. Chap. 6)
- A hazard exposure analysis covering the entire MRS constitutes a positive development, since research has hitherto merely focused on selected municipalities (Vásquez and Romero 2005; Vásquez and Salgado 2009; Ebert et al. 2010)
- The distribution of hazard exposure according to residents' socio-economic status and physical housing conditions

The hypothesis presented in Sect. 7.1 that assumes residents of the MRS will be affected unevenly by climate change as a result of their socio-economic status and their physical housing conditions is therefore analysed in this chapter. The analysis adopts a similar approach to Wisner et al. (2004), who stated that "the poor suffer generally more" (ibid. 13) from the impacts of acute or seasonal natural events. Residents with a lower socio-economic status, for instance, could be left with buildings and plots that are more exposed to hazards as a result of unequal and stratified access to housing and land markets, thus rendering them (physically) more vulnerable to climate change effects.

The analysis furthermore distinguishes between urban and rural populations and housing conditions. In general, it is assumed that "rural residents may be more vulnerable due to lower incomes and more dependent on locally based resource extraction economies" (Cutter et al. 2003: 247). In terms of heat, rural inhabitants are more likely to suffer dry wells as a result of insufficient water acquisition infrastructures and crop losses in the agricultural sector with consequential impact on the rural economy (Cross 2001).

Finally, the exposure of different population groups and dwellings calls for specific adaptation strategies. On the whole, responding to vulnerability with adaptive governance action is an attempt to reduce the vulnerability of natural and human systems against actual or expected climate change effects (Klein et al. 2007). As outlined in Chap. 1, adaptation is a reaction to risks and vulnerabilities of selected key sectors as they occur under climate change. It refers to the adjustment of laws, programmes, plans and measures in an effort to curb negative climate change impacts at urban-regional and municipal level. This chapter takes the exposure of people and dwellings to heat and flood hazard as a starting point for the development of adaptation measures for hazard exposure reduction. Accordingly, it considers the ability to either prepare for exposure to hazards or to cope with or adjust to potential and already traceable negative impacts.

7.3 Methodological Steps: Data, Indicators and Scenarios

In line with the theoretical perspective on vulnerability (Sect. 7.2), socio-economic data is taken into account to identify distributional patterns of hazard exposure among the city's inhabitants. Statistical data on physical housing conditions and the socio-economic features of the MRS population are linked to the land-use analysis findings on areas exposed to flood and heat (cf. Chap. 6) in order to identify specific patterns of exposure.

Due to the scale of the research, which takes the MRS as a whole into account, secondary statistical data is taken from the 2002 Chilean Census of Population and Housing of the Chilean Statistical Institute (INE). It should be remarked that data uncertainty is higher in rural areas, presumably as a result of an increase in the dynamics of urbanization and land-use change, and the ten-year time span between census data updates.

7.3.1 Data Sources

Census data for the MRS is comprised of several spatial resolutions. The blocks (*manzanas*) employed here have the highest resolution. Due to data accessibility issues at this scale, no data is available for a total of nine per cent of the population

(546,468) in the MRS. This applies to all of its 52 municipalities. Hence an analysis of exposure to flood and heat cannot be conducted for this share of 'no data' areas.[1]

7.3.2 Operationalization

Exposure to climate change related hazards is analyzed at different spatial scales (region, municipality and block) for both flood and heat, adopting different time perspectives (status quo 2002 and future scenarios for 2050). The status quo of the spatial distribution of hazard exposure is examined for its relation to socio-economic patterns, employing a methodology designed by Walker et al. (2006) for flood risk in Great Britain. Apart from physical exposure to hazards in determined areas, it is vital to understand who is affected and what characteristics define the population groups exposed to climate change related hazards as a result of their residential location. For this purpose, the population is considered at household level, where their socio-economic status and the physical quality of their housing is taken into account. The methodological steps are presented in the following.

(a) <u>Indices for residents' socio-economic situation and physical housing conditions</u>

Two indices using data from the Chilean census 2002 were applied to analyse the socio-economic status of the exposed residents and their physical housing conditions: the GSE Index and the COFIVI Index. Both serve as a proxy in analysing the socio-spatial distribution of hazard exposure.[2]

- **GSE Index** (Índice de Grupos Socio-Económicos): the GSE Index is widely used in Chile to evaluate the socio-economic situation of households. The Chilean National Census does not comment on income. Consequently the GSE Index refers only to the educational level of the head of household and the possession of ten household goods.[3] These parameter values are ranked and divided into five categories (ABC1, C2, C3, D and E)—with ABC1 referring to the highest socio-economic strata. For our study, the index is calculated for the MRS and the methodology used by Sabatini et al. (2010) and Welz (2012) applied.
- **COFIVI Index** (Índice de COndiciones FÍsicas de VIvienda): the COFIVI Index was created by the authors to describe the physical conditions of housing. It refers to the construction materials used for roofs, walls and floors, as well as to drinking water access and type of sanitation (toilet). It consists of five categories descriptive of housing conditions: I = excellent,

[1] The municipalities of Maria Pinto and San Pedro were excluded from the study, since the data required for an adequate analysis of hazard exposure was unavailable (over 75 % of the population without data at block level).

[2] Other indicators such as age, occupational and employment status are not in the focus of this chapter although they play a key role in assessing vulnerability (Cutter et al. 2003; Lein and Abel 2010).

[3] Colour television, refrigerator, telephone, mobile phone, video recorder, microwave, PC, car, cable television and Internet.

II = superior, III = ordinary, IV = deficient, V = precarious (for further methodological explanations, see Krellenberg et al. 2013). Material and infrastructure rating is based on definitions used by the Chilean Ministry of Housing and Urban Development (MINVU 2006) and the Chilean Statistical Institute (INE).

In a next step, data on socio-economic strata (GSE) and physical housing conditions (COFIVI) is overlaid with maps of hazard exposed areas derived from satellite images (cf. Chap. 6), enabling the identification of hazard exposure by socio-economic situation and physical housing conditions. While a relatively exact analysis of residents' exposure is feasible for flood hazard,[4] there are several methodological constraints on the identification of heat hazard exposure (cf. Chap. 6).

(b) Indicators for the social distribution of hazard exposure

The relative frequency of hazard exposure in relation to socio-economic and physical housing conditions is not the sole focus of our research. In a city like Santiago de Chile, which shows strong socio-spatial differentiation patterns (Kabisch et al. 2012; Rodriguez and Arriagada 2004; Sabatini et al. 2001), the question of uneven distribution also takes prominence. In order to measure the degree of inequality in the distribution of hazard exposure, a modified version of the Walker et al. (2006) 'Comparative Environmental Risk Indicator (CERI)' is employed. Based on a quotient (the ratio of ratios) and calculated for each of the five quintiles of the GSE Index and the COFIVI Index, this indicator is described by the following equation:

$$\textbf{CERI}\,(\text{GSE, COFIVI}) = \frac{\left(\dfrac{\text{Quintile X}^{\text{hazard exposed}}}{\text{Quintile X}}\right)}{\left(\dfrac{\text{N}^{\text{hazard exposed}} - \text{Quintile X}^{\text{hazard exposed}}}{\text{N - Quintile X}}\right)}$$

N = Total Amount
X = Particular Quintile

The quotient is calculated at regional level for both flood and heat exposure, and provides information on the distribution of hazard exposure in the respective quintiles of GSE and COFIVI in relation to other quintiles. The CERI Index supplies information on equity and thus allows for analysis of the likelihood of residents' exposure according to their socio-economic situation and physical housing conditions. The results should be interpreted in combination with hazard frequency and the total number of hazard exposed households and

[4] Residents are exposed to a certain degree when their dwelling or immediate surroundings are flooded. In cases where flood water does not enter the house and there is no material loss, they may nevertheless be unable to leave the building to go to work or to school.

dwellings, since identification of target groups in the context of developing adaptation measures for the MRS is based on these figures.

7.3.3 Scenario Analysis

To allow for conclusions on future exposure and adequate adaptation measures oriented to long-term needs, a scenario analysis of exposure to climate change related hazards in Santiago de Chile is undertaken, following the overall procedure presented in Chap. 3. This is based on:

- The analysis of selected exposure indicators consisting of the number of people exposed to both hazards (flood and heat), and the likelihood of flood and heat hazard exposure differentiated by GSE and COFIVI
- The consideration of two scenarios (storylines) (cf. Chap. 3)
- The incorporation of results obtained from the land-use scenario analysis (cf. Chap. 6) that highlight the dynamics of future flood and heat events under conditions of climate change

In a first step, the progress of the selected exposure indicators is calculated up to the year 2050 under the two scenario alternatives 'Business As Usual' (BAU) and 'Collective Responsibility' (CR) (cf. Chap. 3), taking the dynamics of future hazardous events into account (cf. Chap. 6). In a second step, normative target values for hazard exposure reduction are set for each indicator. The third step consists in exploring whether the target values will be fulfilled under the two scenario alternatives, thus highlighting where adaptation action is urgently required.

7.4 Current Flood and Heat Hazard Exposure in the Metropolitan Region of Santiago

The overall exposure of Santiago de Chile's population to seasonal flooding in the winter and to extreme heat during the summer months is analysed in this section with reference to both hazards. According to the results presented in Chap. 6, 3.7 % (363 km^2) of the surface area of the MRS was exposed in 2002 to frequent flooding at a high level[5] and 0.7 % (70 km^2) to very high surface temperatures (over 35°C). Over and above the physical exposure of determined areas, it is crucial from an adaptation perspective to understand who is exposed and what their socio-economic situation is, as well as the physical condition of their housing. For this purpose, GSE and COFIVI indices are applied (see Sect. 7.3.2). It should be noted that due to the acquisition and processing of census data, the GSE Index is available at household level only, while the COFIVI Index employs information on dwellings. Differentiating an exact estimate of the number of residents exposed to flood and

[5] The official flood maps for Santiago de Chile indicate three levels of flood intensity. High levels of flood exposure refer here to areas flooded at least once every 2 years, as described by Ayala et al. (1987).

heat, respectively, by socio-economic status and physical housing conditions is therefore not possible.[6]

Overall, eleven per cent of the total population was exposed to heat at their place of residence (approx. 666,000 residents), and ten per cent (or 610,000 residents) were exposed to flooding in 2002. Interestingly, flood- and heat-prone areas tend not to overlap, leaving a total of three per cent (approx. 123,000 residents) of the entire MRS population exposed to both hazards. This calls for hazard-specific adaptation measures that target population groups with a heightened likelihood of hazard exposure (above average in comparison to other groups) as well as those with significant numbers of exposed residents.

7.4.1 Flood Exposure

A total of roughly eleven per cent of households (approx. 165,000) and dwellings (approx. 152,000) in the MRS were exposed to floods in 2002. The status quo analysis was unable to identify flood exposure for the urbanized areas[7] of five of the 52 municipalities in the MRS.[8] The share of flood-exposed households is generally lower for the rural areas, where nine per cent of all households (9,833) are exposed, compared with eleven per cent of urban households (154,824). Analysis of the spatial distribution of households exposed to floods at the municipal level confirms this unequal distribution between urban and rural areas. The highest number of flood-exposed households was observed for the urban municipalities of El Bosque (11,783), Ñuñoa (11,710) and Maipú (9,417). In contrast, household flood exposure is very low in the predominantly rural municipalities. This is reflected in the number of flood-exposed dwellings, which shows a similar trend.

In the analysis of flood-exposed households and their socio-economic situation, percentages were highest for the uppermost socio-economic strata (ABC1) (see Fig. 7.1). This distribution is representative for all urban municipalities of the MRS. The opposite is the case for rural areas, where the lower socio-economic strata in particular are exposed to flood hazards, with the D strata making up the highest share at 26 %. Taking into account the total number of flood-exposed households in the whole of the MRS, the D strata (with 55,540 flood-prone households) are exposed most to flooding.

Similar patterns are observed when it comes to the physical housing conditions of flood-exposed dwellings. Dwellings in all types of physical condition were found

[6] An average of 3.7 persons per household and dwelling can be used as a proxy for the MRS in 2002 (INE 2002).

[7] Yet, non-urbanized areas exposed to flooding were identified in some of these municipalities for 2002. As the analysis of land-use change reveals (cf. Chap. 6), urbanization of flood-exposed land is ongoing, more than 3.2 ha of which were urbanized from 2002 to 2009 in the entire MRS. Hence a greater number of inhabitants could in the meantime be exposed to floods.

[8] Alhue, Colina, Padre Hurtado, Peñaflor and San Jose de Maipo, all of which are located in rural environments.

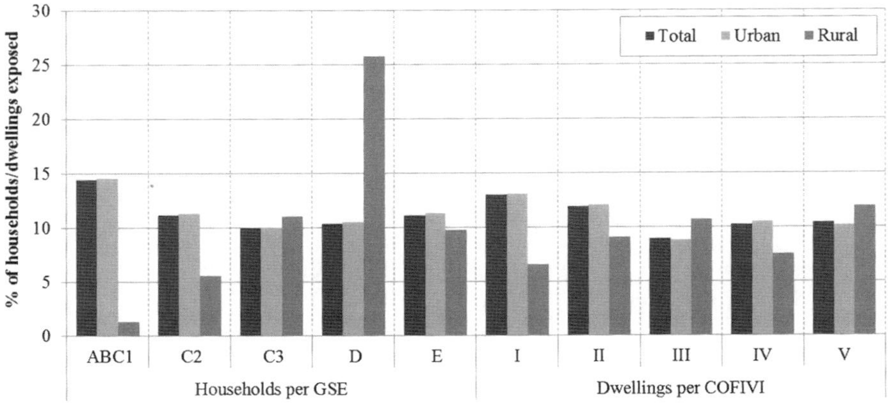

Fig. 7.1 Percentage of households and dwellings exposed to flood, differentiated by urban-rural areas, GSE and COFIVI Index (MRS 2002). (*Source*: Authors' own presentation based on INE (2002))

to be exposed. Of those located in flood-prone areas the largest share was found to be in COFIVI category IV (deficient) with 44,636 dwellings. In relative terms, however, (calculated in relation to the same group), the highest percentages of flood exposure (13 %) are found in the highest COFIVI category I (see Fig. 7.1). This picture is also representative of the urban areas of the MRS, while in rural areas the highest percentages are observed in COFIVI category V with twelve per cent and COFIVI category III with eleven per cent.

A comparison at municipal level shows a high degree of heterogeneity in the physical housing conditions of exposed dwellings in the MRS. Flood-exposed dwellings comprise up to 80 % of dwellings in the most unstable COFIVI category V in some rural municipalities as, for example, Lampa. In contrast, urban municipalities such as Ñuñoa, Vitacura and Las Condes, which have the most stable and most valuable physical housing conditions (COFIVI category I), account for almost 60 % of all flood-exposed dwellings. At the same time, flood mitigation measures recently implemented in upper-scale municipalities such as Providencia and Vitacura indicate a reduction in flood exposure of more prosperous households in the future.

7.4.2 Heat Exposure

Areas exposed to extreme heat (average surface temperatures over 35°C, cf. Chap. 6) are located in the centre as well as in the northern and western parts of the city rather than in the mountainous areas of the eastern and southern parts, where temperatures are considerably lower.[9]

[9] It should be noted, however, that surface temperatures often show quite small-scaled patterns of varying temperatures between adjacent housing blocks.

A total of approx. 14 % of households (approx. 210,000) and dwellings (approx. 190,000) in the MRS were affected by extreme heat in 2002. In general, 14 % of all urban households (189,272) and 19 % of all rural households (20,288) are exposed to extreme heat.[10] With between 13,000 and 27,000 exposed households, the municipalities of Santiago, Pudahuel and Puente Alto are particularly heat exposed. The municipalities with the highest percentages, however, are Lampa with 48 %, followed by Colina, Santiago and Pedro Aguirre Cerda with an approximate 40 % each. It is worth noting that Lampa and Pedro Aguirre Cerda are the only municipalities in the region characterized by high rates of exposure to both flooding and heat.

Exposure to heat in the MRS at the place of residence is clearly associated with low socio-economic status. While 19 % of all households from the E strata and 18 % from the D strata are heat exposed, this rate decreases the higher the households' socio-economic status, with a mere 3 % of all ABC1 households heat exposed (see Fig. 7.2). Hence heat exposure is a problem of both the urban and the rural poor, since D strata households constitute by far the largest group exposed to heat in rural and urban areas.

In light of these findings, dwellings located in heat-exposed areas throughout the MRS are primarily characterized as having precarious, deficient or ordinary housing conditions (see Fig. 7.2). Twenty per cent of all dwellings from the lowest COFIVI category (V) and around 15 % of categories IV and III are heat exposed. In total numbers, COFIVI category IV holds the highest share with approximately 70,000 exposed dwellings.

7.5 Distributional Patterns of Current Hazard Exposure

The analysis of distributional patterns of hazard exposure provides valuable information in the context of target group identification for adaptation policies by describing (a) the characteristics of the exposed population and (b) their specific likelihood to be exposed to one of the examined hazards. For this purpose, the adapted comparative risk indicator (CERI) is used to compare the likelihood of hazard exposure for each individual GSE and COFIVI category (see Sect. 7.3).

Figure 7.3 indicates the likelihood of households in the respective GSE quintile to be exposed to flood or heat compared with households in other quintiles. Figure 7.4 shows the physical housing conditions as measured by the COFIVI Index.

[10] In the status quo analysis, heat exposure could not be identified for seven municipalities in the MRS: Alhue, Calera de Tango, Las Condes, Pircue, San Jose de Maipo, San Pedro and Vitacura, all of which are mountainous locations and/or mainly rural areas. Alhue and San Jose de Maipo are hence the only municipalities in the MRS identified as not exposed to either flood or heat.

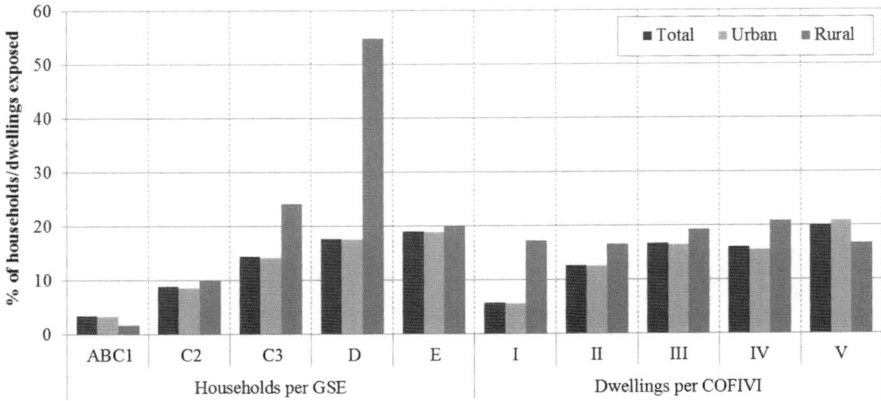

Fig. 7.2 Percentage of households and dwellings exposed to heat, differentiated by urban-rural areas, GSE and COFIVI Index (MRS 2002). (*Source*: Authors' own presentation based on INE (2002))

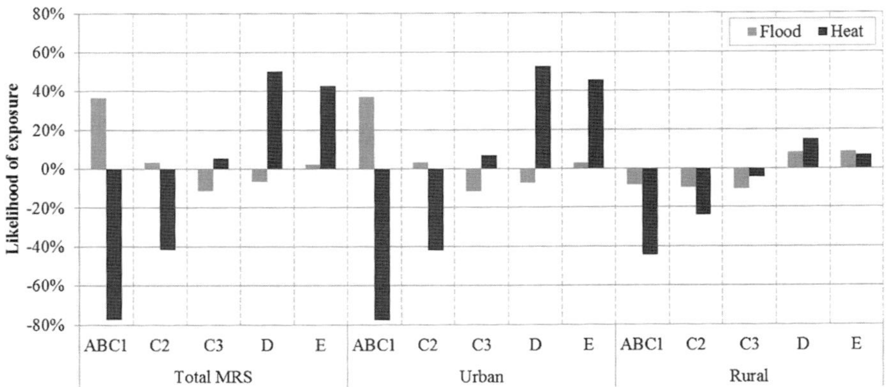

Fig. 7.3 Likelihood of flood and heat exposure by socio-economic conditions (GSE Index) in urban and rural areas, as measured by CERI (MRS 2002). (*Source*: Authors' own presentation)

The analysis of hazard exposure dissimilarity in relation to the households' socio-economic conditions (as measured by the GSE Index) indicates differential trends in flood and heat.

Accordingly, under consideration of exposure likelihood, the issue of flood exposure is associated with the upper socio-economic strata. Households from the ABC1 quintile in the MRS are 37 % more likely to be living in areas exposed to floods than the rest of the MRS population, whereas C3 strata are 11 % and D strata seven per cent less likely to be exposed to this hazard (see Fig. 7.3). Figure 7.3 shows the substantial difference in exposure patterns for urban and rural areas. In the rural areas, the lower socio-economic strata (D and E) in particular are more likely to be flood exposed than the other groups.

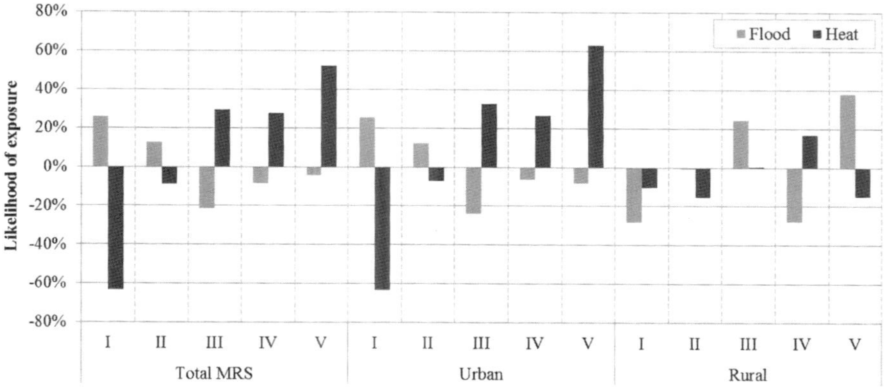

Fig. 7.4 Likelihood of flood and heat exposure by physical housing condition (COFIVI Index) in urban and rural areas, as measured by CERI (MRS 2002). (*Source*: Authors' own presentation)

The correlation between heat exposure and residents' socio-economic status, on the other hand, shows a reverse trend. A distinct pattern of inequality emerges: the lower the socio-economic status, the more likely the exposure to extreme heat at the place of residence. As Fig. 7.3 indicates, households from the D strata are 50 % and those from the E strata 43 % more likely to be exposed to heat, while households with a high socio-economic status are 77 % (ABC1) and 42 % (C2) less likely to be exposed to heat. These results are outstanding in urban areas and lose in significance in the rural areas, where percentages do not reach the same value extremes.

The analysis of hazard exposure dissimilarity in relation to physical housing conditions (as measured by COFIVI, see Fig. 7.4) shows that distributional patterns in many ways reflect the socio-economic stratification of hazard exposure. While dwellings showing better physical housing conditions are much more likely to be exposed to flood than those belonging to other COFIVI categories, the contrary holds true for exposure to heat. Here, dwellings with deficient or precarious housing conditions indicate a very high likelihood of exposure.

A comparison of urban and rural areas within the MRS shows that the above-mentioned distributional patterns of hazard exposure do differ (see Figs. 7.3 and 7.4):

- The urban areas are very close to the general trend, with a tendency towards relatively strong heat exposure of the urban poor (D and E strata) and of dwellings with low to very low physical housing conditions (especially COFIVI category V).
- Households in rural areas from the two lower GSE strata show a heightened likelihood of exposure to both flood and heat. This picture is somehow more diffuse when rural physical housing conditions are considered. In this case, dwellings belonging to COFIVI category III and V are more likely to be flood exposed, while those in category IV are almost 30 % less likely to be exposed to flood hazard but instead more susceptible to heat exposure.

In conclusion, around one-eighth of the population of Santiago across all socio-economic strata and physical housing conditions is exposed to climate change related hazards in one way or another. Flood exposure is more likely among the upper socio-economic strata — as measured at household level (GSE) –, while the occurrence of heat exposure is closely linked to low socio-economic status. Results for physical housing conditions—as measured at dwelling level (COFIVI)—show a similar pattern.

7.6 Hazard Exposure in 2050: Tendencies and Challenges

Looking forward to the year 2050, estimates of future tendencies in urban vulnerability in the MRS are based on statistical calculations and notional assumptions, and accompanied by proposals for context-specific adaptation measures to climate change. Two alternative development paths, Business As Usual (BAU) and Collective Responsibility (CR) are described (see Chap. 3).

The analysis of future trends under the BAU scenario indicates an overall increase in the number of population groups exposed to flood and heat hazards. In terms of flood events and considering the results of the land-use scenario analysis (cf. Chap. 6), greater equality in the likelihood of exposure along GSE and COFIVI categories will not be achieved. For higher socio-economic strata and dwellings with superior conditions, there is an ongoing high probability of flood exposure, while the situation of the middle and lower middle classes along GSE (C3 and D) and COFIVI (III and IV) categories will improve as a result of more progressive protection measures. While the degree of flood exposure among the E strata (GSE) and population groups living in precarious housing conditions (COFIVI category V) is increasing as a result of inadequate social policies and the weak regulation of urban development, the urban vulnerability of these groups will be either sustained or compounded in the future. The land-use scenario analysis likewise indicates a vast increase in the number of areas exposed to heat due to the impact of changing land use and urbanization processes on the nature of the hazard itself (cf. Chap. 6). This scenario shows evidence of a dramatic increase in the number of populations exposed to heat. In a context of growing social inequality, the likelihood of heat exposure of classes in the upper and the lower socio-economic brackets will generally decrease. This also holds true for dwellings of a higher standard. A higher likelihood of exposure might arise for the C3 strata on the GSE Index and the lower COFIVI categories (III, IV and V), which show evidence of growth in the BAU scenario.

Urban planning and environmental protection at municipal and regional level are tightly knit in the CR scenario. New and existent buildings as well as infrastructure at hazard-prone locations are adapted in (urban) design to decrease vulnerability throughout the MRS. Where this is not feasible, the use of hazard-prone areas for residential purposes is prohibited by law. Need-driven involuntary hazard exposure is generally on the wane, since land and housing markets are now more accessible.

This notwithstanding, the number of people exposed to flood and heat hazards is expected to increase further.

In terms of flood hazard, the land-use scenario analysis (cf. Chap. 6) is based on the assumption that the hazard potential for floods remains constant and will decrease in the more densely populated areas in the future. This is due to measures that minimize the negative impact of both urban expansion and climate change. The total number of residents exposed to floods in this scenario is on the rise—albeit to a lesser extent than in the BAU scenario. Concerning the likelihood of exposure in terms of GSE and COFIVI categories, it should be stressed that the highest socio-economic class and the most stable physical housing conditions will not undergo significant change in 2050. On the contrary, the E strata of the GSE Index and COFIVI category V will be less exposed in the CR than in the BAU scenario. The land-use scenario analysis highlights the increase in heat exposure of no more than five per cent as a result of mitigation measures (e.g., roof gardens, solar panels on roofs and maintenance of ventilation corridors) to minimize the negative impact on densely built-up urban environments. Only low-income locations in the western and southern parts of the urban area will experience an increase in negative effects. The key issue in the CR scenario, however, is the growing number of residents exposed to extreme heat. Uneven exposure of residents with differential socio-economic backgrounds will be reduced, so that in social terms exposure will be more uniformly distributed. On the other hand, the likelihood of heat exposure may increase for dwellings with ordinary or deficient material conditions—or at least remain constant. This is due, in particular, to the general improvement in housing conditions under CR.

The key findings of the CR scenario allow for the conclusion that, similar to the BAU scenario, a reduction in the number of exposed residents and dwellings is a possibility. For CR, predictions reveal the obvious need for action to reduce inequalities of hazard exposure in terms of socio-economic status and physical housing conditions. If flood exposure is to be reduced, the focus must be on the upper socio-economic strata and on dwellings in stable physical condition. Heat exposure, on the other hand, calls for action in favour of population groups living under precarious housing conditions, since they will continue to feature the highest exposure.

Hence the concept of reducing the number of exposed residents and dwellings via adaptation measures is gaining in importance. As for flood hazard reduction, a ban on construction in flood-prone areas, the prevention of further imperviousness and an increase in retention areas can lower the number of people exposed to flood events. Moreover, the inequalities of hazard exposure distribution in terms of residents' socio-economic status and the attendant physical housing conditions clearly require action. Emphasis on the lowest socio-economic strata and dwellings in more precarious material conditions will be necessary in order to reduce heat exposure in the MRS. Reducing flood exposure, on the other hand, calls for action in favour of the middle and upper socio-economic strata and better housing conditions—although protection measures aimed at minimizing flood risk have already been applied in some cases.

Recommendations for appropriate anticipatory adaptation measures should nonetheless take coping capacities into account, as well as the issue of susceptibility, which may differ according to socio-economic status and other indicators (Wisner et al. 2004).

7.7 How Might Adaptation be Introduced? Some Recommendations

The previous sections highlighted the population groups in Santiago de Chile most likely to be exposed to climate change related hazards such as flood and heat. While flood hazard is a predominantly 'urban' phenomenon encountered among the better off, extreme heat is a critical issue for the rural and urban poor. This section concentrates on the process of reducing overall exposure and on the locations and groups most affected in the MRS, and recommends specific anticipatory adaptation measures. These measures take into account the ability of the population concerned to either prepare for exposure to hazards and to cope with or adjust to potential and already traceable negative impacts (see Sect. 7.2). There is a consensus that context-specific measures facilitate adaptation to climate change and its impacts (see Smit et al. 1999). The central question is: which context-specific measures should be undertaken to reduce the uneven patterns of exposure to flood and heat hazards and to facilitate adequate adaptation to climate change in the MRS?

Context-specific measures were identified for the MRS using local stakeholder knowledge and experience (cf. Chap. 9). The adaptation needs discussed with the stakeholders were clustered and prioritized to produce a course of action specific to flooding and extreme heat, which was subsequently embedded in the existing policy context. The first set of measures is devoted to expanding the use of biomass by subsidizing green roofs on residential and public buildings, and creating and preserving green spaces, while the second set focuses on installing cool or white roofs to maximize surface albedo. The first set facilitates adaptation to both hazards, the second refers solely to the reduction of elevated indoor temperatures.

7.7.1 Consolidation of Urban Green Spaces

In the context of climate change impacts such as flooding and extreme heat, great importance is attached to the potential of urban green spaces as an adaptation option. Urban green spaces matter independent of their shape, quality, size or soil type. Their many advantages ease the process of adapting cities to climate change. No matter how small, green spaces produce significant cooling effects via evapotranspiration and shade, and act as natural air conditioners (Bowler et al. 2010). In addition they are excellent water managers, as they reduce surface run-off and consequently flooding during peak flows (Chiesura 2004; Gill et al. 2007; Zhang et al. 2012). Careful green space management and maintenance is required, however, in order to avoid excessive absorption of daytime solar energy and night time radiation, as well as extensive watering due to unsuitable vegetation.

Although the number of green spaces per inhabitant increased in the MRS between 2001 and 2009, the total amount of maintained green space is not only well below the WHO guideline of nine m^2 per capita but also unevenly distributed (Hölzl et al. 2011), in turn ushering in an environmental justice issue (Vásquez and Romero 2005). Historically, lack of green spaces in the MRS is a problem of lower-income municipalities: while average green space cover per municipality is 16 %, only five municipalities rank green space cover above 30 % and five other municipalities at less than seven per cent. Furthermore, Reyes and Figueroa (2010) state that green areas in lower-income neighbourhoods are significantly smaller than those in higher-income neighbourhoods despite the existence of a regulatory framework articulating the transfer, construction and maintenance of green spaces.

Against this background, two adaptation measures aimed at reducing exposure to flood and heat are proposed for the MRS: the adoption of parks by, for example, individuals, neighbourhood associations, youth groups, church groups, schools or businesses, and the implementation of green roofs.

In this way members of the community can improve the quality of parks, encouraging them as new 'proprietors' to 'take ownership' of their local park and protect it from vandalism or other misdemeanours. By adopting a park or an individual tree, volunteers commit to maintaining green spaces through tree planting, pruning, clearing, flower-bed preparation and planting. This low-key, informal programme pursues a public-private partnership beneficial to the city as well as to the individuals and groups involved.

Roof and wall planting is likewise an excellent strategy to, firstly, reduce floods by slowing down the flow of rainwater and reduce peak roof run-off and, secondly, to avoid extreme heat by lowering absorption, releasing solar radiation and producing cooling effects via evapotranspiration (Mentens et al. 2006; Stovin 2009). Moreover, green roofs with specially engineered roofing systems on individual buildings (residential, commercial and industrial) can contribute to longer roof life, provide local biodiversity habitats and create aesthetically pleasing architecture (cf. Velazquez 2005). It is, however, important to bear in mind that green roofs are often prohibitively expensive and statically challenging, since older buildings might not be able to support the additional weight. They also require irrigation during the annual dry period. In cities like Santiago, where water scarcity takes centre stage (cf. Chap. 4), roof irrigation can be demanding and should be carefully considered, since its side effects may frustrate the overall benefits.

So far, the architectural design of urban buildings in the MRS has adopted an ecological approach that includes green roofs and the use of environmentally sound materials. The buildings in question are mostly the result of voluntary intervention. In many cases, vegetation on buildings has been used for aesthetic rather than environmental purposes.

7.7.2 Installing Cool or White Roofs

White or cool roofs are flat roofs painted white or surfaced with light or reflective material (e.g., white vinyl) to achieve minimum solar absorption and maximum thermal emission. This helps to "lessen the flow of heat from the roof into the building, reducing the need for space cooling energy in conditioned buildings" (Levinson and Akbari 2010:53).

White roofing costs are comparable to those of conventional roofs. The energy saved with cool roofs, however, results in money saved. The relatively low installation cost of cool or white roofs and the superfluity of strong structures make them appropriate for residential areas. The efficiency and low cost of the measure, its easy application and reduced energy costs make it particularly suited to the target group concerned, i.e., households characterized by a low socio-economic status, and those living under deficient housing conditions.

For the Chilean context, energy efficiency and sustainable and bioclimatic architecture has been widely developed in theoretical terms, also in terms of analysing possible implementation strategies for social housing estates. While the Thermal Regulatory Program—defined by the Ministry of Housing and Urban Development in 1994—includes new building restrictions, it fails to go beyond a handbook of constructive solutions aimed at avoiding thermal loss during the winter months.

7.8 Conclusions

Given the profound patterns of social inequality and ongoing processes of socio-spatial differentiation in Santiago de Chile, it was assumed that residents of the lower socio-economic strata are those most exposed to climate change related hazards. This literature-based hypothesis (Wisner et al. 2004; Satterthwaite et al. 2007) is partially confirmed for the MRS, as the distribution of hazard exposure in the particular context of Santiago de Chile shows class-specific patterns.

Our research on the flood and heat exposure of Santiago's population, however, provides a more sophisticated picture: surprisingly, it is not class alone that makes the difference but also the hazard itself. On the whole, flood and heat exposure observed for the MRS is not confined to households with a specific socio-economic status or dwellings with specific physical conditions but affects all categories reviewed. Those most exposed to hazards are not found exclusively in the lowest socio-economic strata. Rather, a comparison of exposure likelihood showed that the poor in Santiago de Chile are more exposed to heat, while the better-off are more prone to floods. These distributional patterns notwithstanding, a large number of households from the middle and lower socio-economic strata were also found to be exposed to flood hazards, indicating that the need for adaptation action is not solely related to questions of exposure likelihood.

The findings were derived from a comparative analysis of hazard exposure based on the socio-economic circumstances of the exposed population. The application of GSE and COFIVI indices combined with the comparative environmental risk indicator—adapted from Walker et al. (2006) and arising from the context of environmental justice research—were extremely helpful in this realm. Although their application brought valuable information to light for future adaptation strategies, this methodological advancement highlights nonetheless the need for context-specific indicators and spatially differentiated analysis, allowing in turn for more detailed conclusions.

Research on the root causes of class-specific patterns of hazard exposure will be required in the future and could show that some residence locations in hazard exposed areas are the result of deliberate choice for panorama or prestige purposes, with particular reference to middle- and upper-class dwellings.

Furthermore, to adapt to climate change and reduce the vulnerability of the exposed population, it is vital to analyse the susceptibility and coping capacities of the population concerned and to conduct qualitative studies at local level. The answer to the following questions could prove both stimulating and valuable: how well are the exposed households prepared? Do they suffer losses and how do they recover from hazardous events such as seasonal flooding?

Lastly, the findings call for diverse adaptation actions if vulnerability in the MRS is to be reduced. The adaptation measures proposed consider the highly specific patterns of hazard exposure. Clearly these should be in line with overall adaptation strategies, e.g., in education and urban planning (cf. Chap. 9), and integrated into existing urban policies. In view of the current trend to urbanize hazard-exposed locations, Santiago de Chile is likely to experience an increase in the ex-posure of its population to hazard events in the future. Yet this will depend on climate change adaptation policies and on whether urban development does in fact strengthen capacities to protect the hazard exposed population, diminish their vulnerability and, importantly, prevent a surge of new exposures.

References

Adger, W. N., Agrawala, S., Mirza, M. M. Q., Conde, C., O'Brien, K., Pulhin, J., Pulwarty, R., Smit, B., & Takahashi, K. (2007). Assessment of adaptation practices, options, constraints and capacity. In M. L. Parry, O. F. Canziani, J. P. Palutikof, C. E. Hanson, & P. J. van der Linden (Eds.), *Climate change 2007: impacts, adaptation and vulnerability. Contribution of working group II to the Fourth Assessment Report of the Intergovernmental Panel on Climate Change* (pp. 719–743). Cambridge, UK: Cambridge University Press.

Adger, W. N., & Kelly, P. M. (1999). Social vulnexrability to climate change and the architecture of entitlements. *Mitigation and Adaptation Strategies for Global Change, 4*(3–4), 253–266.

Ayala, L., Espinoza, G., & Saragoni, R. (1987). *Estudio de Áreas de Riesgo por Inundación. Segunda Parte*. Santiago de Chile: Ministerio de Vivienda y Urbanismo.

Bowler, D. E., Buyung-Ali, L., Knight, T. M., & Pullin, A. S. (2010). Urban greening to cool towns and cities: A systematic review of the empirical evidence. *Landscape and Urban Planning, 97* (3), 147–155.

Chambers. (1989). Vulnerability, coping and policy. *IDS Bulletin, 20*(2), 1–7.

Chiesura, A. (2004). The role of urban parks for the sustainable city. *Landscape and Urban Planning, 68*(1), 129–138.

Comisión Económica para América Latina y el Caribe (CEPAL) (2011). Anuario estadístico de América Latina y el Caribe, 2011. Santiago de Chile: Naciones Unidas.

Cross, J. A. (2001). Megacities and small towns: Different perspectives on hazard vulnerability. *Environmental Hazards, 3*(1), 63–80.

Cutter, S. L., Boruff, B. J., & Shirley, W. L. (2003). Social vulnerability to environmental hazards. *Social Science Quarterly, 84*(2), 242–261.

da Silva, L., Dias, P., Costa Ribeiro, W., Lima Sant'Anna Neto, J., & Zullo, J., Jr. (2009). *Public policy, mitigation and adaptation to climate change in South America*. Sao Paulo: Instituto de Estudos Avancados da Universidade de Sao Paulo.

INE - Instituto Nacional de Estadísticas (2002). Censo de Población y Vivienda 2002. Santiago de Chile.

Ebert, A., Welz, J., Heinrichs, D., Krellenberg, K., & Hansjürgens, B. (2010). Socio-environmental change and flood risks: The case of Santiago de Chile. *Erdkunde, 64*(4), 303–313.

Fuchs, S., Kuhlicke, C., & Meyer, V. (2011). Editorial for the special issue: Vulnerability to natural hazards – the challenge of integration. *Natural Hazards, 58*(2), 609–619.

Gill, S. E., Handley, J. F., Ennos, A. R., & Pauleit, S. (2007). Adapting cities for climate change: the role of the green infrastructure. *Built Environment, 33*(1), 115–133.

Hidalgo, R. (2004). La vivienda social en Santiago de Chile en la segunda mitad del siglo XX: Actores relevantes y tendencias espaciales. In C. de Mattos & M. E. Ducci (Eds.), *Santiago en la globalización: ¿una nueva ciudad?* (pp. 219–241). Santiago, Chile: SUR Corporación de Estudios Sociales y Educación.

Hölzl, C., Krellenberg, K., Heinrichs, D., Welz, J., Kabisch, S. (2011). How sustainable are processes of social and spatial differentiation in Santiago de Chile? - Current situation and future scenarios for social inclusion. UFZ Discussion Paper **2**/2011. Leipzig.

Kabisch, S., Heinrichs, D., Krellenberg, K., Welz, J., Rodriguez, J., Sabatini, J., & Rasse, A. (2012). Socio-spatial differentiation: Drivers, risks and opportunities. In D. Heinrichs, K. Krellenberg, B. Hansjürgens, & F. Martínez (Eds.), *Risk Habitat Megacity* (pp. 155–181). Heidelberg: Springer.

Klein, R. J. T., Eriksen, S. E. H., Naess, L. O., Hammill, A., Tanner, T. M., Robledo, C., & O'Brien, K. L. (2007). Portfolio screening to support the mainstreaming of adaptation to climate change into development assistance. *Climatic Change, 84*(1), 23–44.

Krellenberg, K., Müller, A., Schwarz, A., Höfer, R., & Welz, J. (2013). Flood and heat hazards in the Metropolitan Region of Santiago de Chile and the socio-economics of exposure. *Applied Geography, 38*(3), 86–95.

Kuhlicke, C., Kabisch, S., Krellenberg, K., & Steinführer, A. (2012). Urban vulnerability under conditions of global environmental change. Conceptual reflections and empirical case examples of growing and shrinking cities. In S. Kabisch, A. Kunath, P. Schweizer-Ries, & A. Steinführer (Eds.), *Vulnerability, Risks and Complexity: Impacts of Global Change on Human Habitats* (pp. 27–38). Göttingen: Hogrefe.

Kuhlicke, C., Scolobig, A., Tapsell, S., Steinführer, A., & de Marchi, B. (2011). Contextualizing social vulnerability: findings from case studies across Europe. *Natural Hazards, 58*(2), 789–810.

Lein, J. K., & Abel, L. E. (2010). Hazard vulnerability assessment: How well does nature follow our rules? *Environmental Hazards, 9*(2), 147–166.

Levinson, R., & Akbari, H. (2010). Potential benefits of cool roofs on commercial buildings: conserving energy, saving money, and reducing emission of greenhouse gases and air pollutants. *Energy Efficiency, 3*(1), 53–109.

Mentens, J., Raes, D., & Hermy, M. (2006). Green roofs as a tool for solving the rainwater runoff problem in the urbanized 21st century? *Landscape and Urban Planning, 77*(3), 217–226.

Reyes, S., & Figueroa, I. (2010). Distribución, superficie y accesibilidad de las áreas verdes en Santiago de Chile. *Revista EURE, 36*(109), 89–110.

Ribot, J. C., Najam, A., & Watson, G. (2009). Climate variation, vulnerability and sustainable development in the semi-arid tropics. In E. Lisa, F. Schipper, & I. Burton (Eds.), *Adaptation to climate change* (pp. 117–160). London: Earthscan.

Rodriguez, J. (2001). Segregación residencial socioeconómica: ¿qué es?, ¿cómo se mide?, ¿qué está pasando?, ¿importa? Santiago de Chile: CEPAL / CELADE – SERIE Población y Desarrollo N° 16.

Rodriguez, J., & Arriagada, C. (2004). Segregación Residencial en la Ciudad Latinoamericana. *Revista EURE, 30*(89), 5–24.

Sabatini, F., Cáceres, G., & Cerda, J. (2001). Segregación residencial en las principales ciudades chilenas: Tendencias de las tres últimas décadas y posibles cursos de acción. *Revista EURE, 27* (82), 21–42.

Sabatini, F., & Salcedo, R. (2007). Gated communities and the poor in Santiago, Chile: Functional and symbolic integration in a context of aggressive capitalist colonization of lower-class areas. *Housing Policy Debate, 18*(3), 577–606.

Sabatini, F., Salcedo, R., Wormald, G., & Cáceres, G. (Eds.). (2010). *Tendencias de la Segregación en las Principales Ciudades Chilenas*. Santiago: INE.

Satterthwaite, D., Huq, S., Pelling, M., Reid, H., & Lankao, P. R. (2007). *Adapting to climate change in urban areas. The possibilities and constraints in low- and middle-income nations*. London: International Institute for Environment and Development.

Smit, B., Burton, I., Klein, R. J. T., & Street, R. (1999). The science of adaptation: A framework for assessment. *Mitigation and Adaptation Strategies for Global Change, 4*(3–4), 199–213.

Stovin, V. (2009). The potential of green roofs to manage urban stormwater. *Water and Environment Journal, 24*(3), 192–199.

UN-Habitat (2012). Estado de las ciudades de América Latina y el Caribe 2012. Rumbo a una nueva transición urbana. Nairobi, UN-Habitat

MINVU - Ministerio de Vivienda y Urbanismo (2006). Atlas de la evolución del déficit habitacional en Chile 1992-2002. Santiago de Chile.

Vásquez, A., & Romero, H. (2005). Evaluación ambiental del proceso de urbanización de las cuencas del piedemonte andino de Santiago de Chile. *Revista EURE, 31*(94), 97–117.

Vásquez, A., & Salgado, M. (2009). Desigualdades socioeconómicas y distribución inequitativa de los riesgos ambientales en las comunas de Peñalolén y San Pedro de la Paz. Una perspectiva de justicia ambiental. *Revista de Geografía Norte Grande, 43*, 95–110.

Velazquez, L. S. (2005). Organic green roof architecture: Sustainable design for the new millennium. *Environmental Quality Management, 14*(4), 73–85.

Walker, G., Burningham, K., Fielding, J., Smith, G., Thrush, D., & Fay, H. (2006). *Addressing environmental inequalities. Flood risk*. Bristol: Environment Agency.

Welz, J. (2012). Segregation und Integration in Santiago de Chile zwischen Tradition und Umbruch. Zur Bedeutung von sozioökonomisch heterogen strukturierten Stadtquartieren und Mechanismen der Sozialintegration für die (Re-)Produktion von Segregation. PhD Thesis. University of Hamburg.

Wisner, B., Blaikie, P., Cannon, T., & Davis, I. (2004). *At risk. Natural hazards, people's vulnerability and disasters*. London: Routledge.

Zhang, B., Xie, G., Zhang, C., & Zhang, J. (2012). The economic benefits of rainwater-runoff reduction by urban green spaces: A case study in Beijing, China. *Journal of Environmental Management, 100*, 65–71.

Synthesis: Climate Change Impacts from a Cross-Sectoral Perspective: Consequences for Political Response

8

Kerstin Krellenberg

Abstract

The foregoing Chaps. 4, 5, 6 and 7 have indicated the need for action in response to the pressing issues of land-use change, its linkage to natural hazards, the associated vulnerabilities, and the supply of and demand for water and energy in the Metropolitan Region of Santiago de Chile. It was evidenced that serious problems already exist in this context and will most likely be aggravated under future climate change conditions. Consideration of other driving factors such as demographic, socio-economic and technological change in the assessment of future development paths allows for the assumption that sectoral developments are interlinked, e.g., land-use change and hazard generation or energy and water supplies. This chapter takes a closer look at these interlinkages and discusses them in the light of cross-sectoral policy recommendations.

Keywords

Climate change impacts • Interlinking processes • Cross-sectoral policies

8.1 Climate Change: Why a Cross-Sectoral Perspective Matters

Climate change is a complex phenomenon and the result of numerous interwoven—physical, socio-economic, social and technological—processes (e.g., Smith and Stern 2011), all of which calls for a comprehensive climate change impact assessment analysis. This complexity is particularly evident when it comes to cities, where urbanization, changing land use and soaring greenhouse gas (GHG)

K. Krellenberg (✉)
Department of Urban and Environmental Sociology, Helmholtz Centre for Environmental Research – UFZ, Permoser Str. 15, 04318 Leipzig, Germany
e-mail: kerstin.krellenberg@ufz.de

K. Krellenberg and B. Hansjürgens (eds.), *Climate Adaptation Santiago*,
DOI 10.1007/978-3-642-39103-3_8,

emissions are in the process of altering the climate (cf. Chap. 1). Consequently, climate change cannot be seen as an isolated process.

Response strategies to the complexity of climate change and its associated impacts necessitate an integrative policy approach. Here the strong sectoral interlinkage, e.g., between water, energy, planning and housing, demands cross-sectoral interaction, not least when adaptation comes to the fore. This means that adequate climate change (adaptation) response must take into account the interaction between policies in different areas and between policies at different levels of decision-making (Adger et al. 2005; Bulkeley 2010; Ohlson et al. 2005; Urwin and Jordan 2008). Interaction should take place horizontally across different policy sectors and vertically across different levels of government (e.g., Corfee-Morlot et al. 2009), rendering it a multi-level governance concern. And it is especially the multi-level complexities of climate change adaptation that make integration challenging (Huntjens et al. 2012; Adger et al. 2005).

Furthermore, climate change-related policy measures should consider synergies with existing measures (e.g., Burton et al. 2002; Corfee-Morlot et al. 2011; Dovers 2009; Smith and Stern 2011), e.g., those emerging from natural resource or water management, disaster risk management, urban planning, sustainable development, or poverty reduction (Füssel 2007). In this context, elements of existing institutional frameworks are often identified as obstacles to climate change response action, e.g., weak and illegitimate governance and institutional structures, lack of coordination between administration levels and rejection of change (cf. Burch 2010; Measham et al. 2011; Sánchez-Rodríguez et al. 2008). In other words climate change response challenges institutional dynamics, since it goes beyond traditional and heterogeneous local administrative structures and capacities to respond (e.g., Huntjens et al. 2012). Changes to institutions and governance will be needed, however, if strategies for adaptation to climate change are to be developed and successfully put into effect.

Calling attention to the results of Chaps. 4, 5, 6 and 7, this chapter points out the interlinkages of several ongoing processes that drive development in the MRS and their capacity for cross-sectoral adaptive policy response action. Section 8.2 synthesizes the different climate change impacts, while Sect. 8.3 discusses the cross-sectoral adaptation measures designed to bridge the sector divide.

8.2 Climate Change Impacts and Their Interlinkages in the MRS

The authors of Chap. 4 have highlighted the close relationship between precipitation and the water supply, on the one hand, and the inflow of surface water from the Andes, on the other. As precipitation occurs for the most part in the winter, inter-annual water supply variations are high and in the summer depend heavily on melt water from glaciers and snowfields. The authors stress that in so-called "dry" years the demand for water already exceeds the supply. Furthermore, water demand in the MRS is characterized by the conflict between various water users. Changes in temperature, precipitation and stream flow predicted for the future will exacerbate current problems, as less water will be available than required. The increase in water demand arising from a growing population and added irrigation can only be compensated in part by new technologies. In other words the water sector is strongly determined by climate conditions

In the energy case, the authors of Chap. 5 have identified a strong interlinkage between ongoing urbanization and the energy sector. Persistent population and GDP growth rates have increased the demand for energy in the MRS. Higher temperatures will lead to a further increase in the demand for electricity to serve air conditioning and refrigeration installations, and a decrease in the consumption of gas and petroleum products used for heating. Although half of the electricity available in the MRS is generated by hydropower (and the other half by thermo-electric power), climate change will affect the energy sector only slightly. In this context, Chile's dependency on fossil fuels and the highly concentrated market (Simon et al. 2012) are expected to cause major problems.

The authors of Chap. 6 have demonstrated the close link between ongoing urbanization patterns, including increased sealed surfaces, and the generation of flood and heat hazards, and their aggravation through climate change. The loss of green areas and agriculturally used land is a conspicuous factor in the reduction of the key cooling and retention functions crucial to flood and heat hazard prevention. The authors of Chap. 7 have shown the increase in exposure of people and housing to both hazards.

As these selected "sectors" indicate, none can be regarded in isolation when it comes to assessing climate change impacts and more particularly to developing climate change response. Water is crucial and therefore inextricably bound up with all other issues. Given the predicted decline in the supply of water due to less annual precipitation and higher temperatures (cf. Chap. 2), water is becoming a scarce commodity, which in turn simultaneously affects both water and energy supplies. Although more green areas are needed to cool the city and serve as retention areas for temporary intensive rainfall, they require additional irrigation most of the time, as the MRS is located in a dry climate with few winter rainfall events. Importantly, ongoing expansion of the urban area has led to a dramatic reduction in green space.

This means that the interwovenness of the processes climate change involves is also an issue for the MRS, while the processes themselves call for cross-sectoral responses, since interventions in one sector can produce positive or negative consequences in others.

8.3 How to Develop Non-Sectoral Adaptation Measures

Designing interventions to combat the cross-sectoral challenges, inherent in climate change, calls for careful consideration of interlinkages. The links between mitigation and adaptation should not be ignored but rather considered simultaneously and as interconnected (cf. Füssel 2007; IPCC 2011; Lowe and Lorenzoni 2007; Tompkins and Adger 2005). Climate change response, notably in cities of the global south, should be closely linked to development strategies (cf. Huq et al. 2006;

Sánchez-Rodríguez et al. 2008). Approaching both issues as interconnected could foster mutually positive effects and increase the potential success of development and climate change policies (OECD 2009; Huq et al. 2006). Furthermore, adaptive response action is tightly linked to disaster risk reduction strategies, since both aim at reducing vulnerability by enhancing the capacity to anticipate, resist and recover from hazards. Wherever possible, therefore, any synergies between them should be exploited (cf. OECD 2009).

Again, the distribution of competences at different levels of decision-making, the connections and interdependencies these levels entail, and the resultant need for coordination call for an integrative approach in the case of cross-sectoral adaptation measures. Participation could foster the inclusion of a wide range of actors from different sectors and administrative levels. In a participatory process, stakeholders must be integrated from the outset of the project to enable cross-sectoral consideration of the interwoven processes and impacts involved. As numerous scholars have stated, scientists, practitioners, decision-makers, policy analysts, local communities, NGOs, private sector representatives, government agencies, and international organizations, among others, need to be involved to guarantee an integrative policy-making process (cf. Füssel 2007; Krellenberg 2012; OECD 2009; Watson 2005). Here, a scientifically organized participatory process has the potential to bring actors and knowledge together in a more 'neutral' environment (Krellenberg 2012; Krellenberg & Barth 2014).

On the other hand, the participatory process does not guarantee per se the effective implementation of cross-sectoral measures. To give a concrete example: the development of climate change response action for the MRS (Chap. 9) led to the 14 context-specific adaptation measures presented in Chaps. 4, 5, 6 and 7. Given the impact interlinkage described in Sect. 8.2 of this chapter, the proposed adaptive response measures are of necessity also interlinked and cross-sectoral. Careful implementation will thus be required to achieve the intended positive effects. Prior to implementation, co-benefits and the potential impact of each measure on other sectors should be assessed in order to arrive at a 'win-win situation' rather than maladaptation.

The example of green space makes the relationship between the water and the energy sector palpable. Developing green areas as a response to increasing heat and flood hazard is of the utmost importance. Given that water is now more scarce in the MRS and irrigation demands are on the increase, this dilemma could be solved in part by reusing grey water. Possible conflicts of interest between green areas and the energy sector could arise against the backdrop of the surface available for solar and wind installations and the attempt to diversify energy sources.

It is important that decision-makers are aware of the interlinkages. Hence these issues are emphasized and discussed in the present example of the Regional Climate Change Adaptation Plan for Santiago de Chile. Criteria and metrics were defined and included in the Plan to facilitate evaluation of the adaptation measures and their success within the given policy environment (cf. Moser 2009). This and other issues are presented and discussed in the next chapter.

References

Adger, W. N., Arnell, N. W., & Tompkins, E. L. (2005). Successful adaptation to climate change across scales. *Global Environmental Change, 15*, 77–86.

Bulkeley, H. (2010). Climate policy and governance: An editorial essay, Wiley Interdisciplinary Reviews: Climate Change, pp.1311–313.

Burch, S. (2010). Transforming barriers into enablers of action on climate change: Insights from three municipal case studies in British Columbia, Canada. *Global Environmental Change, 20*, 287–297.

Burton, I., Saleemul, H., Lim, B., Pilifoscva, O., & Schipper, E. L. (2002). From impacts assessment to adaptation priorities: The shaping of adaptation policy. *Climate Policy, 2*, 145–159.

Corfee-Morlot, J., Cochran, I., Hallegatte, S., & Teasdale, P. J. (2011). Multilevel risk governance and urban adaptation policy. *Climatic Change, 104*, 169–197.

Corfee-Morlot, J., Kamal-Chaoui, L., Donovan, M. G., Cochran, I., Robert, A., Teasdale, P.-J. (2009). Cities, Climate Change and Multilevel Governance, OECD Environmental Working Papers14, OECD Publishing.

Dovers, S. (2009). Normalizing adaptation. *Global Environmental Change, 19*, 4–6.

Füssel, H. M. (2007). Adaptation planning for climate change: concepts, assessment approaches, and key lessons. *Sustainability Science, 2*, 265–275.

Huntjens, P., Lebel, L., Pahl-Wostl, C., Camkin, J., Schulze, R., & Kranz, N. (2012). Institutional design propositions for the governance of adaptation to climate change in the water sector. *Global Environmental Change, 22*, 67–81.

Huq, S., Reid, H., & Murray, L. A. (2006). Climate change and development links. *Gatekeeper Series, 123*, 3–19.

IPCC (2011). Special Report. Managing the Risks of Extreme Events and Disasters to Advance Climate Change Adaptation (SREX), available online at: http://ipcc-wg2.gov/SREX/. Published 18th November 2011.

Krellenberg, K. (2012). A science-policy approach towards local adaptation planning: The case of Santiago de Chile. In: Otto-Zimmermann, K. (ed.) Resilient Cities 2: Cities and Adaptation to Climate Change - Proceedings of the Global Forum 2011 (Local Sustainability).

Krellenberg, K., & Barth, K. (2014). Inter- and transdisciplinary research for planning climate change adaptation responses—the example of Santiago de Chile *Interdisciplinary Science Reviews, 39*.

Lowe, T. D., & Lorenzoni, I. (2007). Danger is all around: Eliciting expert perceptions for managing climate change through a mental models approach. *Environmental Change, 17*, 131–146.

Measham, T. G., Preston, B. L., Smith, T. F. Brooke, C., Gorddard, R., Withycombe, G., & Morrison, C. (2011). Adapting to climate change through local municipal planning: barriers and challenges. *Mitigation Adaptation Strategies Global Change, 16*, 889–909.

Moser, S. (2009). Making a difference on the ground: the challenge of demonstrating the effectiveness of decision support. *Climatic Change, 95*, 11–21.

OECD (2009). Integrating climate change adaptation into development co-operation. Policy guidance. Available at: http://www.oecd.org/dataoecd/45/45/44887764.pdf.

Ohlson, D.W., McKinnon, G.A. & Hirsch, K.G. (2005). A structured decision-making approach to climate change adaptation in the forest sector. In: The Forestry Chronicle, January-February 2005, (Vol. 81, no.1, pp. 97–103).

Sánchez-Rodríguez, R., Fragkias, M. & Solecki, W. (Lead Authors) (2008). Urban responses to climate change: a focus on the Americas. Workshop Report. June 2008. Available online at: ccsl.iccip.net/ur2cc.pdf. Accessed 21 February 2013.

Simon, S., Stelzer, V., Vargas, L., Paredes, G., Quintero, A. (2012). Energy systems. In: Heinrichs et al., Risk Habitat Megacity, pp. 183–205

Smith, L. A., & Stern, N. (2011). Uncertainty in science and its role in climate policy. *Philosophical Transactions of The Royal Society, 369*, 4818–4841.

Tompkins, E. L., & Adger, W. N. (2005). Defining response capacity to enhance climate change policy. *Environmental Science & Policy, 8*, 562–571.

Urwin, K., & Jordan, A. (2008). Does public policy support or undermine climate change adaptation? Exploring policy interplay across different scales of governance. *Global Environmental Change, 18*, 180–191.

Watson, R. T. (2005). Turning science into policy: challenges and experiences from the science-policy interface. *Philosophical Transactions of The Royal Society, 360*, 471–477.

Part IV

Adaptation Strategy: Developing Measures and Implementation

Developing Climate Change Adaptation Measures in a Participatory Process: Roundtable Meetings

9

Jonathan Barton, Jordan Harris, and Kerstin Krellenberg

Abstract

Participation plays a major role in contemporary urban planning. This is particularly the case for climate change adaptation, which—given the interwoven processes of climate change—involves a wide range of actors and sectors. This chapter discusses the overall need for participatory adaptation planning, exemplifying it with the process to develop a Regional Climate Change Adaptation Plan for Santiago de Chile. The experience highlights the most significant lessons learned, including the challenges and constraints that emerged during the process. The chapter stresses the use of a multi-stakeholder, inter-sectoral planning approach that involved the organization of ten roundtable meetings in Santiago de Chile over a period of two and a half years. Political legitimacy was provided by the two principal institutions responsible for climate change planning at city-regional level. Their participation in the process from the outset was central to the successful elaboration of the Regional Climate Change Adaptation Plan.

Keywords

Urban adaptation measures • Participatory adaptation planning • Roundtable meetings

J. Barton (✉) • J. Harris
Instituto de Estudios Urbanos y Territoriales, Pontificia Universidad Católica de Chile, El Comendador 1916 Providencia, Santiago de Chile, Chile
e-mail: jbarton@uc.cl

K. Krellenberg
Department of Urban and Environmental Sociology, Helmholtz Centre for Environmental Research – UFZ, Permoser Str. 15, 04318 Leipzig, Germany

K. Krellenberg and B. Hansjürgens (eds.), *Climate Adaptation Santiago*,
DOI 10.1007/978-3-642-39103-3_9, © Springer-Verlag Berlin Heidelberg 2014

9.1 Introduction: Participatory Adaptation Planning

Since its origins in the 1985 Villach meeting organized by United Nations Environment Programme (UNEP) and the World Meteorological Organization (WHO), the international debate on climate change has firmly established itself in the scientific community. Proving the anthropological contribution to climate change of greenhouse gas emissions was not an easy task given the opposition generated by the fossil fuel lobby, particularly in the USA. Consequently, the onus has been on science research to unmistakeably identify anthropogenic connections as a prerequisite for public policy action. The Nobel Prize awarded to Al Gore and the Intergovernmental Panel on Climate Change (IPCC) scientists in 2007 went some way towards establishing consensus on this human influence. Public perceptions, however, are not always in tune with science. While societal ambivalence towards the reduction of greenhouse gas emissions (mitigation) may not be that problematic, since action focuses on power plants and heavy industry, when it comes to adaptation the situation is more complex.

Climate change planning has emerged alongside a general awareness that planning should be a more participatory process. Healey (1992) refers to it as 'planning through communication' as distinct from the less flexible, expert-led planning that prevailed in the 1960s and 1970s. It is precisely this actor-oriented type of planning that has become key to the discussion on adaptation to climate change (Eisenack and Stecker 2011a, b; McLaughlin and Dietz 2007; Nelson et al. 2007).

This chapter describes the participatory planning process to generate a Regional Climate Change Adaptation Plan for Santiago de Chile containing context-specific adaptation measures. It sees participation as a wholly political process (Arnstein 1969), where decision-making is influenced by a wide variety of actors. Giddens (1998) stresses the role of diverse 'stakeholders' as active political agents who can and should be more directly involved in decisions that affect their lives. The push towards participatory planning in general and participatory adaptation planning in particular should be understood in this context.

The first question to be posed here is who should participate and with what legitimacy. In other words, what criteria should be employed to assemble a group that sit at a roundtable meeting and contribute to the construction and design of a climate change adaptation plan? Secondly, how can the process be organized to ensure a balanced exchange of science and policy, to embrace local needs and knowledge, and to finally arrive at feasible adaptation measures?

Conspicuous in the literature on climate change adaptation planning is its action-oriented nature (Eisenack and Stecker 2011a, b; Miller et al. 2010). Apart from producing strategic planning documents to define areas for intervention and responsibilities, it is crucial to ground them in specific measures. The latter can either be 'soft', e.g., education and communication, or 'hard', e.g., infrastructure development (Sovacool 2011). This is where participation plays a vital role, for example in consensus building and the legitimizing of concrete measures, generating pressure for finance and awareness in society as a whole (Burton et al. 2002; Corfee-Morlot et al. 2011; Moser 2009). Discussions on the different governance frameworks in the interests of effective adaptation planning are inherent in the participatory process. It is paramount to recognize the importance of private sector and non-governmental actors in the generation and execution of the measures in question. Participation complements rather than replaces a governmental, public-led approach, opening up the process to other actors and sectors in order to enhance transparency, achieve more effective planning, and make the appropriate links across different actor and interest groups (Ansell and Gash 2007; Few et al. 2007; Tompkins and Adger 2003). A participatory process is not a 'given', however. It has to be defined, and its limitations and barriers recognized. It is precisely the positive elements and constraints that arise in the participatory planning process experience for adaptation that are the focus of this chapter.

The process of developing adaptation measures addressed in this book follows the participatory planning patterns of Healey (1992). Not only is there a strong recognition of the links between science and policy, and the need to establish that interface clearly in order to convert scientific information into practical measures. There is likewise an acceptance that the process of creating adaptation measures is not one of 'truth' (science) simply talking to 'power' (executive decision-makers) in a one-way system where political actors are obliged solely to act on information provided by scientists; instead, the dynamic of the process requires a two-way 'communication exercise'. The aim of this two-way system is to reinforce information generation by combining scientific 'facts' with the practical expert knowledge of decision-makers, and thus further the efficacy of the measures concerned (Fankhauser et al. 2007; Mitchel et al. 2006; Roux et al. 2006; Vogel et al. 2007). Recognition of the need to convert scientific knowledge, where available, into concrete actions is evident for decision-making (Corfee-Morlot et al. 2011; Dovers 2009; Prato 2008; Tschakert and Dietrich 2010) and emphasized by the IPCC as well as by the national authorities.

The participatory process presented in this chapter was organized as a series of roundtable meetings for the purpose of designing adaptive measures at the science-policy interface to offset the adverse effects of climate change. It involved governmental and non-governmental representatives from different sectors and administrative levels in discussions on the scientific information on climate change and its impacts, institutional responsibilities, and alternative measures for adaptation. To highlight the distinctive elements of this process, which led to a range of context-specific adaptation measures, the chapter is structured as follows: Section 9.2 provides an overview of the participatory planning process to generate adaptation measures; Section 9.3 discusses some aspects of the political process involved; Section 9.4 gives information on the 14 adaptation measures developed for the Regional Climate Change Adaptation Plan; Section 9.5 discusses some of the difficulties on the way to reaching consensus in the course of the participatory planning process; Section 9.6 concentrates on the importance of participation for adaptation planning; Section 9.7 draws a general conclusion.

9.2 The Participatory Planning Process to Develop Adaptation Measures: A Description of People, Products and Events

Several multi-lateral institutions have recognized the significance of identifying ongoing adaptation processes and their integration into formal policies, plans and investments (OECD 2009; Stern 2006; UNFCCC 2007; World Bank 2010). In this way, the end result of the presented participatory planning process is a Regional Climate Change Adaptation Plan for the Metropolitan Region of Santiago (MRS), with 14 - context-specific adaptation measures to counteract the impacts that climate change implies for the Metropolitan Region of Santiago de Chile (MRS). The Plan, which contains key findings on the effects of climate change, builds on the input, discussions and participation of the numerous stakeholders involved in the process and on the interaction between stakeholders, scientists and experts. Each of the 14 measures was subjected to an in-depth scoping study to determine and systematize the legal, financial and institutional elements to be considered prior to implementation of the measures. Data from the studies was included in the Regional Climate Change Adaptation Plan and the complementary Manual for Implementation, which serves as a guide for public officials from the various agencies responsible for putting the measures into practice.

For a better understanding of the participatory process to produce context-specific adaptation measures for the MRS, this section highlights how it was organized and who was involved.

The process adopted a multi-stakeholder, inter-sectoral planning approach. It involved the organization of ten roundtable meetings held over a period of two and a half years with key institutional representatives (primarily technical rather than political) from a broad spectrum of social sectors, and policy- and decision-makers in a joint endeavour to work out a set of concrete adaptation measures for the region (Fig. 9.1). The aim was to bring together a group of leading actors who would remain involved in the process throughout its duration, creating a cohesive and collaborative environment where information could be shared and views, experiences and opinions exchanged in the interests of meeting the main adaptation needs of the MRS as presented in Chaps. 4, 5, 6 and 7 of this book. Furthermore, the process was designed to support and, to some extent, legitimize scientific work in political and societal communities. In this sense, the approach exemplifies anticipatory, planned, public adaptation to climate change (IPCC 2001; Krellenberg 2012).

The entire process received its 'legitimacy' from the strong involvement of and collaboration with the two main political actors of climate change in the MRS, the Regional Government (GORE) and the Regional Secretariat of the Ministry of the Environment (SEREMI MA). It not only entailed creating measures for adaptation to climate change, but also their grounding in the form of a Regional Climate Change Adaptation Plan to be implemented by the pertinent actors in the public sector. Despite the support of the two principal political actors, there was no binding requirement for representatives from other ministries, services, private firms or NGOs to engage in the participatory process. In this sense, a fragile process had to be driven by the facilitators in order to maintain interest in it and generate the Plan as the ultimate objective.

Successful adaptation planning to climate change calls for strong support from the local decision-making structure in the public sector (Barton 2009; Corfee-Morlot et al. 2011; Dovers 2009; Smit and Wandel 2006). Hence an interdisciplinary group of scientists worked closely with representatives from GORE and SEREMI MA to organize the process presented in this chapter. Working partners were identified in both institutions. Maintaining them, however, posed a challenge due to (1) shifting job positions of the political and technical authorities within the public system, and (2) the metamorphosis of environmental institutionality in Chile

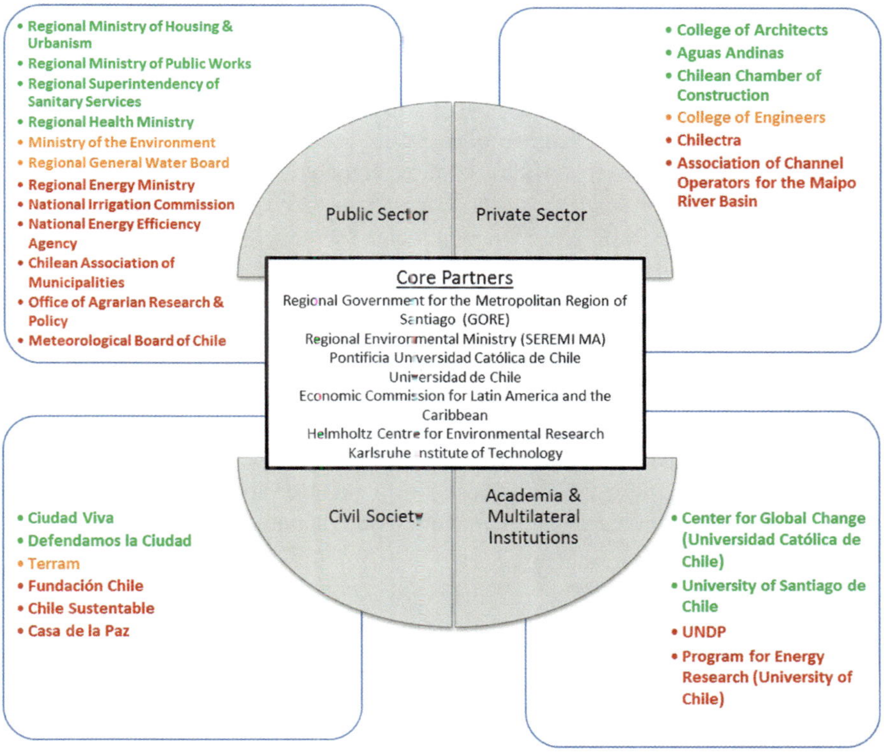

Fig. 9.1 Overview of participating institutions at the roundtable meetings (*Source*: Authors' own presentation)
Note: Organizations written in green attended all or most of the roundtable meetings from the beginning of the process. Those written in orange displayed a medium level of participation, i.e., were absent from several roundtable meetings. Organizations written in red took part in one or two roundtable meetings only

during the initial phase of the process. The National Commission of the Environment (CONAMA), which had initially been identified as a working partner, was decommissioned in 2010 and replaced by the newly established Ministry of the Environment (MMA). This entirely new environmental institutionality in the Chilean government complicated the formalizing of collaboration with the newly inaugurated MMA to the extent already in place with GORE. The issue of shifting job positions was a permanent challenge throughout the participatory process, as key participants of the roundtable meetings were either replaced or dropped out. This notwithstanding, the high interest level of both key partner institutions sustained their involvement. Subsequent meetings with GORE led to solidification of the Regional Governor's support for and commitment to the creation of adaptation measures in an international science-policy research process, formalized with the signing of a Memorandum of Understanding. A professional from the Planning and Development Division became one of the key partners for the duration of the entire process, greatly facilitating the continuing participation and support of the Regional Government. SEREMI MA followed the same procedure.

Another vital aspect of planning is the formation of a common framework of concepts and goals by local actors from public and private sectors, civil society and scientific and policy experts, enabling them to establish a process of sharing information and reaching consensus on policy issues and initiatives (Shaw et al. 2007; UNDP 2010; OECD 2009), and to explore the possible effects of the expected climate change on a variety of stakeholders involved in local governance. The process was designed as multi-sectoral and participatory, since the involvement of representatives from a variety of sectors has been shown to increase the legitimacy, effectiveness and viability of adaptation planning and measures (Jones et al. 2009; Shaw et al. 2007; Vogel et al. 2007). To establish the core group of between 15 and 20 permanent members representing the relevant organizations from the individual sectors, a list of potential stakeholders from the public and private sectors, civil society, academia and multi-lateral institutions was drawn up in cooperation with GORE. Throughout the roundtable meeting process the frequency of participation by these representatives and their overall commitment to the process varied considerably. Additional participants were invited to attend when specific issues associated with the main climate change impacts and the corresponding adaptation measures were discussed (Chaps. 4, 5, 6 and 7). As shown in Fig. 9.1, a number of institutions were present throughout the entire process.

The importance of securing a science-policy interface in climate change adaptation planning came to the fore in the last decade (Adger et al. 2005; Bulkeley 2010; Birkmann et al. 2010; Corfee-Morlot et al. 2011; Füssel 2007; Tompkins and Adger 2005). Some authors even state the need to provide useful scientific information for decision-makers in an interactive participatory process (Mitchel et al. 2006; Vogel et al. 2007) in order to create awareness of adaptation needs and generate adequate responses built on effectiveness, efficiency, equity and legitimacy (Adger et al. 2005; Pelling 2011), while at the same time taking into account social, cultural, political and institutional constraints based on the concept of adaptive capacity (Smit and Pilifosova 2001). Hence all of the roundtable meetings included presentations by scientists or experts, followed by group discussions and participatory activities.

The roundtable meeting was structured in a number of stages with scientific and stakeholder input. The first phase introduced the issue of climate change and adaptation, providing strong scientific input to inform stakeholders of the predicted impacts of climate change on the MRS (Chap. 2), of the expected local impacts (Chaps. 4, 5, 6 and 7), and of the importance of adaptation in general. This procedure was complemented by information from the participants on their current activities and responsibilities related to climate change and its impacts, and their efforts to determine current measures and distinguish between adaptation and mitigation. This stakeholder input was crucial to establishing a common contextual framework that would allow the issues at hand to be tackled to maximum effect (c.f. Krellenberg & Barth (2014)).

The next phase explored adaptation planning and measures, the use of scenarios in the context of climate change planning, and international examples of adaptation plans and measures. Based on the scientific input, stakeholder actions were contextualized against a backdrop of other examples of adaptation planning. This was followed by a debate on the findings of scientific studies on the implications of climate change impacts on the MRS, and the proposal of policy responses and concrete measures appropriate to this context. Based on the scientific input, the final phase saw prioritization of the measures following stakeholder participation and input. This led to in-depth discussions of the legal, financial and institutional needs and implications for the implementation of the measures (c.f. Krellenberg & Barth (2014)).

As transparency and a common understanding are two vital components of successful participatory processes (Few et al. 2007; Collins and Ison 2009), the information derived from discussions and participatory activities was systematized and sent out to participants in the form of roundtable minutes after each meeting. The information gleaned from one activity served as input for participatory activities at the next roundtable. Roundtable meetings included discussion papers produced by the scientists involved on a number of issues associated with climate change adaptation, and a summary of the key findings on climate change and its impacts on the MRS. These documents served to create a common knowledge base for roundtable participants, enabling them to make informed contributions to the discussions on adaptation measures and the implementation of the final outcome, the Regional Climate Change Adaptation Plan. Figure 9.2 lists the issues, activities and documents corresponding to roundtable meeting.

9.3 Political Process and the Politics of Participation

Apart from developing adaptation measures in a participatory process, gaining political support for the final implementation of the adaptation plan is indispensable (Burton et al. 2002; Lim et al. 2005; Corfee-Morlot et al. 2011). In the present case of the Regional Climate Change Adaptation Plan for the MRS and parallel to the roundtable process, efforts were made to garner the political support needed to socialize and pass the Plan by the various institutional actors involved in regional planning and policy-making. The majority of stakeholders in the roundtable process were technical professionals from various institutions and agencies, making it necessary to likewise involve the political operatives in charge of these institutions. Two interesting aspects arise: where does responsibility lie for (a) pushing the process and (b) including the maximum number of pertinent actors?

A new Regional Governor took office in the middle of the planning process. Much of the latter's success in the political sphere was due to his continued interest and support, which led to the creation of new political spaces within the Regional Government, including the establishment of a Sub-Committee on Climate Change. In addition, the politically appointed Regional Ministerial Secretary also proved to be a staunch supporter. Consequently the process achieved a high level of acceptance. The final products were presented to the Regional Council, the political body responsible for assigning funds and approving regional initiatives. Uninterrupted close relations with the Office of Climate Change in the MMA led to the

	Scientific Presentations	Participatory Activities	Briefing Papers
08/2010	RT 1: • Introduction to the Project: Adaptation and Regional Urban Planning	RT 1 • Introduction of the participants and incorporation of activities related to climate change	
11/2010	RT 2 • Preliminary results of expected climate change impacts on the MRS	RT 2 • Identification of issues and interrelation of the areas of interest: water, energy, land use and vulnerability	BP 1 • Urban Adaptation to Climate Change: Issues and Impacts
01/2011	RT 3 • Final results for expected climate change impacts on the MRS	RT 3 • Identification of existing participant activities and interrelation of the four key areas of interest	
04/2011			BP 2 • Urban Adaptation to Climate Change: The role of planning in adaptation
05/2011	RT 4 • Scenarios and adaptation planning: Visions for the future of the MRS	RT 4 • Evaluation of existing activities relevant to climate change adaptation and long-term strengths and weaknesses	BP 3 • Scenarios and Urban Planning BP 4 • Urban Adaptation Plans: Cases and experiences
09/2011	RT5 • Land use, threats and vulnerabilities in the MRS: Adaptation measure proposals	RT 5 • Discussion and proposal of concrete adaptation measures to diminish exposure to threats	BP5 • Adaptation Planning Instruments Report • Land Use and Vulnerability Summary Report
11/2011	RT 6 • Water resources in the MRS: Adaptation measure proposals for water management, supply and demand	RT 6 • Discussion and proposal of concrete adaptation measures	BP 6 • Risk and Vulnerability in Adaptation Planning Report • Water Summary Report
01/2012	RT 7 • Energy in the MRS: Adaptation measure proposals for energy supply and demand	RT 7 • Discussion & proposal of concrete adaptation measures	Report • Energy Summary Report
04/2012	RT 8 • Presentation and Evaluation of Climate Change Adaptation Measures	RT 8 • Prioritizing measures according to financial, administrative and institutional criteria	• Regional Climate Change Adaptation Plan for the MRS
06/2012	RT 9 • Political, administrative and institutional needs for implementation of the Regional Climate Change Adaptation Plan	RT 9 • Establishing responsibilities and roles in the implementation of the Regional Climate Change Adaptation Plan	BP 7 • Inter-Sectoral and Spatial Planning for Climate Change Adaptation Measures

Fig. 9.2 Roundtable structure and content. (*Source*: Authors' own presentation)

institutionalizing of the Regional Climate Change Adaptation Plan within the framework of the Chilean National Climate Change Action Plan, as the only regional plan to be accepted in a series of sectoral plans that are being generated between 2012 and 2015. Thus the overall process gained tremendous recognition from a variety of key political agents in regional and national structures that converge in the MRS. It has been handed over to GORE and SEREMI MA, the institutions responsible for implementing the Plan.

9.4 Development of Adaptation Measures

Adaptation measures must involve institutional and individual responses, some of which are included in ongoing programmes and processes (Adger et al. 2007:721). Initiatives already in operation should therefore be analysed for their potential influence (in any direction) on adaptation goals, to improve coordination and to deal with the risk of maladaptation.

The 14 context-specific adaptation measures that ultimately entered the Regional Climate Change Adaptation Plan were the result of the participatory planning process described in Sect. 9.2. These measures were elaborated as a response to the adaptation needs presented in Chaps. 4, 5, 6 and 7. Existing measures and activities were taken into account, based on input provided by local stakeholders. The need for action is conspicuous, as climate change and parallel processes are expected in the future to aggravate the problems that currently prevail in the sectors examined in the impact analysis. Designing short-term responses with a long-term perspective to reduce the identified adverse effects of climate change seems essential. An overview of all 14 adaptation measures is presented in the following.

Adaptation Measures: Land Use: Heat and Flood Hazard Reduction

Measure 1: Monitoring system—WebGIS

Measure 2: Green factor in new housing (public or commercial)

Measure 3: Use of existent irrigation channels along the Andean pediment to reduce flood risk

Adaptation Measures: Vulnerability: Reduction of flood and heat hazard exposure

Measure 4: Programme for the implementation of green roofs

Measure 5: Public participation in maintenance and creation of green areas

Measure 6: Technical programme for passive cooling in low-income housing

Adaptation Measures: Water

Measure 7: Reduction of water demand through low-budget sanitary installations in existent housing and hotels

Measure 8: Awareness-raising of water treatment and re-use of grey water in new housing

Measure 9: Reduction of water demand in agriculture through new irrigation technologies

Measure 10: Implementation of water management structure for the Maipo/Mapocho catchment

Adaptation Measures: Energy

Measure 11: Formation of public energy groups in GORE MR and themunicipalities

Measure 12: Education on climate change and energy

Measure 13: Diversification of energy supply sources

Measure 14: Reduction of domestic energy consumption

Smit et al. (1999) distinguish between different categories of adaptive measures. Differences refer to specific sectors and the necessary adaptation to the impacts involved, as well as to variations in the spatial (national, regional and local) and temporal (short-, medium-and long-term) scales. The 14 measures selected for the MRS reflect what stakeholders identified as the most relevant topics, which were initially selected for their overall objectives and suitability to local conditions. The more explicit the measures are formulated, the more easily they can be implemented. They were summarized and prioritized according to their likelihood of reaching the desired target and practical implementation, their suitability to the local context, previous experience with them, and the balance between temporal and spatial scopes.

Converting adaptation measures into action calls for close scrutiny of institutional frameworks. In many cases, regulatory and structural/operational elements are identified as barriers to action when it comes to adapting to climate change. These include weak and illegitimate governance and institutional structures, lack of coordination between administrative levels, and resistance to change (cf. Burch 2010; Measham et al. 2011; Sanchez-Rodriguez et al. 2008). Hence elements such as a supportive institutional and policy environment at national and international level, interdepartmental cooperation, as well as consistency of adaptation programmes and those dealing with non-climatic issues are deemed necessary for successful local adaptation to climate change (cf. Burch 2010, cf. Chap. 8 of this book).

On the other hand, adaptation to climate change challenges institutional dynamics, since it goes beyond traditional and heterogeneous local impacts and response capacities (cf. Huntjens et al. 2012). Changes in institutions and in governance are thus indispensable to the development and effective implementation of climate change adaptation strategies. Adger et al. (2009) emphasize the importance of smooth governance given the diversity of actors involved and the potential divergence of interests, both elements that could paralyse action. Consequently the measures selected for the MRS were analysed by local experts to determine their applicability to the existing institutional, legal, and financial context. The scoping studies identified several obstacles. Some of the selected measures propose regulatory or institutional changes, while others suggest technological solutions. The findings were submitted to a feedback loop via one of the roundtable meetings, where the necessary adjustments were made.

For measures dealing with climate change to be meaningful, their long-term effects should be established and both previous and current conditions and experiences in the planning and implementation stages taken into consideration (cf. Hildén 2011). The final selection of 14 adaptation measures took into account the temporal horizon reaches from short to long-term, and the spatial scope reaches from the national level of basic institutional or regulatory changes to site-specific structural solutions. It is likewise vital to successful implementation that information be provided on expected benefits, previous implementation experience, and tangible practical solutions that include the required changes to the institutional and legal framework of Chile and the MRS.

9.5 The Complexities of Consensus

The plans to develop concrete adaptation measures for the Regional Climate Change Adaptation Plan for Santiago envisaged a participatory process that would forge an interface between the science and policy fields. The creation of an intermediate space for technical experts and civil society representatives to discuss experiences, existing positions and projects, and viable options for implementation constituted a platform to ground the measures in the local context. Although the consensus-building experience is often post hoc and consultative (Cooke and Kothari 2001; Owens et al. 2004), here the desire to breed consensus around the adaptation measures became a key component of the participatory process. The formation of roundtables allowed participants to contribute throughout the process and—it can be assumed—formed the backdrop to full agreement on the final plan, albeit not without heated debate on further additions (extending and deepening the range of measures) during the penultimate working roundtable.

There can be little doubt that participation in policy-making is a minefield, in terms of who should participate, their legitimacy, and the relationship between technical and political inputs. Participation is generally considered a prerequisite for effective adaptation strategy implementation, transmission of information, enhancement of adaptive capacities, active listening to and embracing of different viewpoints and different interests in an effort to avoid problems of legitimacy and stagnation, and to collect pertinent local knowledge (cf. Tompkins and Adger 2005; Burch 2010; Adger et al. 2007; Adger et al. 2009; Eisenack and Stecker 2011a, b; Sanchez-Rodriguez et al. 2008).

The abuse of participatory processes is normally related to post-hoc consultations (Cooke and Kothari 2001). In the context of this process, the specific modality of the roundtables in the MRS constituted the first of two participatory steps. The second was a public consultation process, the phases and timing of which were defined by Chilean law. Passing on the Plan to public consultation was promoted by the Climate Change Department of the Chilean Ministry of Environment. The creation of two opportunities for participation on the path from science to implementation showed a recognition of participation as having many forms and is not exclusive to any one of them. The main challenge is how to integrate responses and ensure that subsequent steps based on these inputs are transparent, bearing in mind that due to their diversity it is not feasible to integrate all responses in equal measure, e.g., those that call for a cancellation or freezing of the plan. It is this transparency that is frequently lacking in participatory processes

(Rydin and Pennington 2000), hence the fatigue that is generated time and again. In the roundtable process under review, which became the backbone of the Regional Climate Change Adaptation Plan for the MRS, the space for contributions, criticism of the documentation and the format, and alternative proposals was considerable. The small scale of the roundtable format makes it potentially more manageable—depending on the goodwill of the participants—compared to the public consultation exercise (expected during the first semester of 2013).

As highlighted by Few et al. (2007), Keeney and McDaniels (2001), and Pielke (1998) among others, recognizing the political complexities involved in decision-making processes related to climate change is crucial. In this regard, the process of garnering political consensus for the Regional Climate Change Adaptation Plan for the MRS was not without obstacles. On several occasions, political opposition emerged between members of the Regional Council and the administrators from GORE, threatening the advance of the process. As climate change is an emerging issue in the regional governance apparatus in the MRS, a number of politicians questioned its relevance and importance, and blocked opportunities to socialize the Plan further with political decision-makers. At one point, the supervisor of the GORE representative disallowed this key partner's participation in the remainder of the process, jeopardizing the possibility of maintaining the relevance of the Plan within this body. At each of these junctures, the dominant factors that contributed to upholding the process were the close coordination and support between the technical and political operatives of GORE (including the Governor's office, the President of the Sub-Committee on Climate Change, and the professional from the Planning and Development Division), SEREMI MA (including the Regional Secretary and the head of the Natural Resources, Biodiversity and Climate Change Department), and MMA (including the head of the Office of Climate Change and other professionals working there).

9.6 The Sine Qua Non of Participation in Climate Change Adaptation

Adaptation to climate change is closely related to a wide range of more generalized development issues, including food security, housing and services provision, infrastructure and socio-economic status. Given this proximity to the development agenda, especially the issue of risk assessment and management, there are difficulties in stressing the specific nature of adaptation, above and beyond other factors. For example, what are the additional risks generated by climate change for those living in vulnerable spontaneous settlements located in areas designated as 'risk-prone'? Since these factors are difficult to establish and part of the uncertainty of climate change sciences (both natural and social), there is a need to apply the precautionary principle laid down in the Rio Declaration of 1992. This principle promotes the logic of 'risk-averse' decision-making in the face of uncertainty or lack of scientific evidence. It is precisely for these reasons—the difficulty of proving the additional contributions of climate change to existing development conditions, and the precautionary principle—that participation in adaptation is key. A further consideration vital to adaptation planning is the influence of context specificities. Adaptation is highly context specific and requires bottom-up responses that reflect both micro- and meso-scale conditions relating to cultural, socio-economic, physical and productive dimensions.

In the same way that development discourse since the 1960s has shifted from expert-led to community-led planning, and that urban and regional planning has adopted a similar shift in thinking, the need to mesh scientific research and actor-oriented processes and create a science-policy interface becomes a sine qua non of climate change adaptation (Adger et al. 2009; Healey 1992). In the case of the presented approach, this interface was established through a roundtable process that lasted approx. two and a half years. While the obstacles to effective participation remain a concern, e.g., legitimacy or the binding nature of participation, it is no longer appropriate to conceive of adaptation planning without the creation of these spaces for dialogue and exchange. Science in and of itself is not a reason to change practices and habitual forms of 'doing development'. What is required is a process of communication for planning that uncovers the fragility of the science, as well as its strengths, and which also overlaps with existing actions, demands and awareness.

The roundtable meetings for developing the Regional Climate Change Adaptation Plan for the MRS provided good examples of the range of discussion points that emerge in relation to climate change adaptation in urban areas. Links to watershed management, concerns over current and future land-use plans, contrasting agendas across government institutions, and the lack of time-series data for a number of relevant variables were among these points. By creating a space to connect science to these issues, the measures that were generated were able to establish possible 'sticking points' for implementation and the search for alternatives. It was also possible to document existing investments and programmes that relate to adaptation, although more implicitly or tangentially rather than explicitly.

9.7 Conclusions

The process of developing adaptation measures for the MRS was designed to bring together scientific modelling exercises and the participation of a large number of stakeholder organizations. The 'soft science' of participatory planning—based on a fusion of 'hard science', different perceptions of regional conditions, political and technical interests, and a broad spectrum of expectations—was instituted as a means of generating legitimacy for the final Plan. Despite the legitimacy provided by the two principal partners within the region, GORE and SEREMI MA, implementation and acceptance depend on a wider range of actors. 'Opening up' the discussion over the options and opportunities of adaptation planning was the specific objective of the roundtable process.

The 14 measures that constitute the final Plan are products of a science-policy interface mediated by the significant number of stakeholders that participated over the approx. two-and-a-half-year period. The length of time required and the political or technical appropriateness or legitimacy of each of these participants remain points open to discussion. It was not, however, the object of the exercise to create a single model or perfect set of conditions for participation.

Most importantly, the participants of the roundtable meetings were representatives of different institutions and organizations with a role in adaptation planning as promoters, implementers or critics. By opening up the potential fields of conflict or disagreement within a defined process, the measures could be honed and concerns minimized. The resultant Plan, although not without potential obstacles in the context of a public consultation exercise or even within the government itself at regional and central levels, was effectively legitimized as a consequence. This is not a plan that emerged in final form from a university department or government office, but one that was forged over a long period of time and "socialized" in the process.

Participation is a mixture of the formal and the informal. It brings actors to the table (in the belief that their participation will enhance the outcome) who may benefit or indeed endanger the process, it may founder on political decisions taken during or after the process, and it may generate more doubts rather than less. It is precisely the uncertainty of participation that facilitates unexpected outcomes or the pursuit of new lines of inquiry or concern. Conventional science and the positivist method seek to reduce uncertainty and provide 'solid' evidence.

Each participatory process has its own dynamic. Given that it is a political process, there is no single formula that can guarantee a particular outcome. A balance has to be struck between active debate and consensus building. Clearly the role of information and communication is key to this process, as is the opportunity to share ideas and to listen to alternatives. For scientists, this is not always the most obvious way forward. It is, however, this process of iteration, of adapting to the contributions of participants, and of generating a fusion of project-based science and alternative knowledge (based on other scientific projects, on experience, or on credible assumptions) that is part of this 'post-normal' methodology. By post-normal we refer to a post-positivist approach to knowledge creation and sharing whereby expertise is compared and contrasted with the non-expert knowledge of diverse participants. The goal of such an approach is to challenge certain scientific assumptions, to challenge expertise as the only valid knowledge source, and to lead to enhanced sharing of different viewpoints and constructions to create a more viable outcome.

Urban development can rarely be reduced to a small number of variables that are easily controlled. Urban systems are defined by their complexity. Consequently, complexity, uncertainty and the dynamism of participation have to be considered as part and parcel of adaptation. In this particular case, participation at the science-policy interface was successful, albeit final political approval of the Regional Climate Change Adaptation Plan for the MRS must be taken outside the boundaries of the roundtable meeting process. While the experience described is not the only means of conducting a participatory process related to climate change adaptation planning, it is a valid example transferable to other urban areas.

References

Adger, W. N., Agrawala, S., Mirza, M. M. Q., Conde, C., O'Brien, K., Pulhin, J., Pulwarty, R., Smit, B., & Takahashi, K. (2007). Assessment of adaptation practices, options, constraints and capacity. In M. L. Parry, O. F. Canziani, J. P. Palutikof, P. J. van der Linden, C. E. Hanson, M. L. Parry, O. F. Canziani, J. P. Palutikof, P. J. van der Linden, & C. E. Hanson (Eds.), *Climate Change 2007: Impacts, Adaptation and Vulnerability. Contribution of Working Group II to the Fourth Assessment Report of the Intergovernmental Panel on Climate Change* (pp. 717–743). Cambridge: Cambridge University Press.

Adger, W. N., Arnell, N. W., & Tompkins, E. L. (2005). Successful adaptation to climate change across scales. *Global Environmental Change, 15*, 77–86.

Adger, N. W., Dessai, S., Goulden, M., Hulme, M., Lorenzoni, I., Nelson, D. R., Naess, L. O., Wolf, J., & Wreford, A. (2009). Are there social limits to adaptation to climate change? *Climatic Change, 93*, 335–354.

Ansell, C., & Gash, A. (2007). Collaborative governance in theory and practice. *Journal of Public Administration Research and Theory, 18*(4), 543–571.

Arnstein, S. R. (1969). A ladder of citizen participation. *Journal of the American Institute of Planners, 35*, 216–224.

Barton, J. R. (2009). Adaptación al cambio climático en la planificación de ciudades-regiones. *Revista de Geografía Norte Grande, 43*, 5–30.

Birkmann, J., Garschagen, M., Kraas, F., & Quang, N. (2010). Adaptive urban governance: new challenges for the second generation of urban adaptation strategies to climate change. *Sustainability Science, 5*(2), 185–206.

Bulkeley, H. (2010). Climate policy and governance: an editorial essay. *Wiley Interdisciplinary Reviews: Climate Change, 1*, 311–313.

Burch, S. (2010). Transforming barriers into enablers of action on climate change: insights from three municipal case studies in British Columbia, Canada. *Global Environmental Change, 20*, 287–197.

Burton, I., Huq, S., Lim, B., Pilifosova, O., & Schipper, E. L. (2002). From impacts assessment to adaptation priorities: the shaping of adaptation policy. *Climate Policy, 2*, 145–159.

Collins, K., & Ison, R. (2009). Jumping off Arnstein's ladder: social learning as a new policy paradigm for climate change adaptation. *Environmental Policy and Governance, 19*, 358–373.

Cooke, B., & Kothari, U. (Eds.). (2001). *Participation: the new tyranny?* New York: Zed.

Corfee-Morlot, J., Cochran, I., Hallegatte, S., & Teasdale, P.-J. (2011). Multilevel risk governance and urban adaptation policy. *Climatic Change, 104*, 169–197.

Dovers, S. (2009). Normalizing adaptation. *Global Environmental Change, 19*, 4–6.

Eisenack, K., & Stecker, R. (2011a). *An Action Theory of Adaptation to Climate Change. Earth System Governance Working Paper No. 13*. Amsterdam: Earth System Governance Project.

Eisenack, K., Stecker, R. (2011b). A framework for analyzing climate change adaptations as actions. Manuscript appears in mitigation and adaptation strategies to global change. doi:10.1007/s11027-011-9323-9.

Fankhauser, S., Smith, J., & Tol, R. (2007). Weathering climate change: some simple rules to guide adaptation decisions. *Ecological Economics, 30*, 67–78.

Few, R., Brown, K., & Tompkins, E. L. (2007) Public participation and climate change adaptation: avoiding the illusion of inclusion. *Climate Policy, 7*(1).

Füssel, H.-M. (2007). Adaptation planning for climate change: concepts, assessment approaches, and key lessons. *Sustainability Science, 2*, 265–275.

Giddens, A. (1998). *The third way: the renewal of social democracy*. Blackwell: Oxford.

Healey, P. (1992). Planning through debate: the communicative turn in planning theory. *Town Planning Review, 63*(2), 143–162.

Hildén, M. (2011). The evolution of climate policies – the role of learning and evaluations". *Journal of Cleaner Production, 19*, 1798–1811.

Huntjens, P., Lebel, L., Pahl-Wostl, C., Camkin, J., Schulze, R., & Kranz, N. (2012). Institutional design propositions for the governance of adaptation to climate change in the water sector. *Global Environmental Change, 22*(1), 67–81.

IPCC - Intergovernmental Panel on Climate Change. (2001). *Climate change 2001: impacts, adaptation, and vulnerability, summary for policymakers and technical summary of the working group II report*. Geneva: IPCC.

Jones, H., Jones, N., Walker, D., Walsh, C., Jones, H., Jones, N., Walker, D., & Walsh, C. (2009). *Strengthening science–policy dialogue in developing countries: a priority for climate change adaptation*. London: Overseas Development Institute.

Keeney, R. L., & McDaniels, T. L. (2001). A framework to guide thinking and analysis regarding climate change policies. *Risk Analysis, 21*, 989–1000.

Krellenberg, K., (2012). A science-policy approach towards local adaptation planning: The case of Santiago de Chile. In Otto-Zimmermann, K. (Ed.) *Resilient Cities 2: cities and adaptation to climate change - proceedings of the global forum 2011 (local sustainability)*.

Krellenberg, K., & Barth, K. (2014). Inter- and transdisciplinary research for planning climate change adaptation responses—the example of Santiago de Chile *Interdisciplinary Science Reviews, 39*.

Lim, B., Spanger-Siegfried, E., Burton, I., Malone, E., & Huq, S. (Eds.). (2005). *Adaptation policy frameworks for climate change: developing strategies, policies and measures*. Cambridge: Cambridge University Press.

McLaughlin, P., & Dietz, T. (2007). Structure, agency and environment: toward an integrated perspective on vulnerability. *Global Environmental Change, 39*(4), 99–111.

Measham, T., Preston, B. L., Smith, T. F., Brooke, C., Gorddard, R., Withycombe, G., & Morrison, C. (2011). Adapting to climate change through local municipal planning: barriers and challenges. *Mitigation and Adaptation Strategies for Global Change, 16*, 889–909.

Miller, F., Osbahr, H., Boyd, E., Thomalla, F., Bharwani, S., Ziervogel, G., Walker, B., Birkmann, J., Van der Leeuw, S., Rockström, J., Hinkel, J., Downing, T., Folke, C., & Nelson, D. (2010). Resilience and vulnerability: complementary or conflicting concepts? *Ecology and Society, 15* (3), 11. [online] URL: http://www.ecologyandsociety.org/vol15/iss3/art11/.

Mitchel, R. B., Clark, W. C., Cash, D. W., & Dickson, N. (Eds.). (2006). *Global environmental assessments: information, institutions, and influence*. Cambridge: MIT.

Moser, S. (2009). Making a difference on the ground: the challenge of demonstrating the effectiveness of decision support. *Climatic Change, 95*, 11–21.

Nelson, D. R., Adger, N., & Brown, K. (2007). Adaptation to environmental change: contributions of a resilience framework. *The Annual Review of Environment and Resources, 32*, 395–419.

OECD. (2009). *Integrating climate change adaptation into development co-operation: policy guidance*. Paris: OECD.

Owens, S., Rayner, T., & Bina, O. (2004). New agendas for appraisal: reflections on theory, practice, and research. *Environment and Planning A, 36*, 1943–1959.

Pelling, M. (2011). *Adaptation to climate change: from resilience to transformation*. London: Routledge.

Pielke, R. A., Jr. (1998). Rethinking the role of adaptation in climate policy. *Global Environmental Change, 8*, 159–170.

Prato, T. (2008). Accounting for risk and uncertainty in determining preferred strategies for adapting to future climate change. *Mitigation and Adaptation Strategies for Global Change, 13*, 47–60.

Roux, D., Rogers, K., Biggs, H., Ashton, P., & Sergeant, A. (2006). Bridging the science-management divide: moving from unidirectional knowledge transfer to knowledge interface and sharing. *Ecology and Society, 11*(1), 4.

Rydin, Y., & Pennington, M. (2000). Public participation and local environmental planning: the collective action problem and the potential of social capital. *Local Environment, 5*(2), 153–169.

Sanchez-Rodriguez, R., Fragkias, M., Solecki, W., (2008). 'Urban responses to climate change a focus on the Americas: a workshop report' urbanization and global environmental change project, an IHDP core project.

Shaw, R., Colley, M., & Connell, R. (2007). *Climate change adaptation by design: a guide for sustainable communities*. London: TCPA.

Smit, B., Burton, I., Klein, R.J.T., & Wandel, J. (1999). *The anatomy of adaptation to climate change and variability. Climatic Change*. Hollinwood, Oldha: RAP Spiderweb Ltd.

Smit, B., & Pilifosova, O. (2001). Adaptation to climate change in the context of sustainable development and equity. In IPPC (Ed.), *Climate change 2001: impacts, adaptation and vulnerability*. Cambridge: Cambridge University Press.

Smit, B., & Wandel, J. (2006). Adaptation, adaptive capacity and vulnerability. *Global Environmental Change, 16*, 282–292.

Sovacool, B. K. (2011). Hard and soft paths for climate change adaptation. *Climate Policy, 11* (2011), 1177–1183.

Stern, N. (2006). *The economics of climate change: the Stern review*. Cambridge: Cambridge University Press.

Tompkins, E., Adger, W.N., (2003). Building resilience to climate change through adaptive management of natural resources. Tyndall Centre for Climate Change Research Working Paper 27.

Tompkins, E., & Adger, N. (2005). Defining response capacity to enhance climate change policy. *Environmental Science and Policy, 8*(6), 562–571.

Tschakert, P., & Dietrich, K. (2010). Anticipatory learning for climate change adaptation and resilience. *Ecology and Society, 15*(2), 11.

UNDP (United Nations Development Programme). (2010). *A toolkit for designing climate change adaptation initiatives*. New York: UNDP.

United Nations Framework Convention on Climate Change (UNFCCC) (2007). *Investment and Financial Flows to Address Climate Change*. Geneva.

Vogel, C., Moser, S. C., Kasperson, R. E., & Dabelko, G. (2007). Linking vulnerability, adaptation, and resilience science to practice: pathways, players, and partnerships. *Global Environmental Change, 17*, 249–364.

World Bank. (2010). *The economics of adaptation to climate change: a synthesis report*. Washington D.C: World Bank.

From Planning to Implementation: Capacities and Competences

10

Kerstin Krellenberg, Katrin Barth, and Natàlia García Soler

Abstract

This chapter discusses the opportunities and constraints involved in implementing adaptation measures under real-world conditions. Based on an assessment of the participatory process that cornerstoned the elaboration of the Regional Climate Change Adaptation Plan for Santiago de Chile, it examines the execution of the 14 measures contained in the Plan. The discussion highlights the challenges along capacities and competences, and takes as its starting point a review of the literature, scientific assumptions and an evaluation by the actors who were part of the particpatory process described in Chap. 9. Although it will remain hypothetical until such time as the measures have been implemented under real-world conditions, the discussion helps to fill the general knowledge gap on the implementation of climate change adaptation measures.

Keywords

Capacities • Climate change adaptation measures • Competences • Implementation

K. Krellenberg (✉) • K. Barth
Department of Urban and Environmental Sociology, Helmholtz Centre for Environmental Research – UFZ, Permoser Str. 15, 04318 Leipzig, Germany
e-mail: kerstin.krellenberg@ufz.de

N. García Soler
Natàlia García Soler, Martinstraße 7, 36433 Moorgrund, Germany

Natàlia García Soler (scholarship provided by Foundation La Caixa—German Academic Exchange Service (DAAD), 2010–2012).

K. Krellenberg and B. Hansjürgens (eds.), *Climate Adaptation Santiago*,
DOI 10.1007/978-3-642-39103-3_10,

10.1 From Theory to Practice: Challenges to the Implementation of Climate Change Adaptation

Numerous cities across the globe have begun to respond to climate change (see, e.g., Heinrichs et al. 2013; Kumar 2013; Krellenberg et al. 2013) but the ability to design and implement local policy responses that engender effective action has, in general, yet to emerge (Burch and Robinson 2007; Huq et al. 2006). According to Adger et al. (2005), a broad distinction can be drawn between climate change action that involves policies or regulations to build adaptive capacity and action related to the implementation of operational adaptation decisions. In other words implementation of climate change response takes place at different levels, in different ways and at different times. Accordingly, the term implementation referred to in this book distinguishes three dimensions or phases in the adaptation process:

1. Implementation of the process of developing an adaptation plan or strategy with the attendant measures.
2. Implementation of the adaptation plan or strategy in the legal and institutional framework.
3. Implementation of individual measures included in the plan or strategy.

Ad (1): Applying this to the concrete example of Santiago de Chile, the first dimension refers to the participatory process of developing the Regional Climate Change Adaptation Plan and its 14 individual measures as described in detail in Chap. 9. It concerns the achievements and constraints associated with organizing and establishing this process at the interface between science and policy/practice (cf. Chap. 1, Fig. 1.1, see Krellenberg and Barth 2014).

Ad (2): The second dimension, implementation of the Regional Climate Change Adaptation Plan and its 14 measures, is highly political in nature and refers to the embedment of the Plan in the legal and institutional framework, e.g., existing regional and national policies, programmes, strategies and plans. Its success depends primarily on political will, on leadership and on aspects such as legal frameworks and public consultancy, all of which determine the veracity of implementation as a genuine political act. Some of the activities undertaken to urge political entities in this direction have been described in Chap. 9 of this book, including efforts to garner political support, presentation of the final Plan to the Regional Council, provision of information required for the Plan to be handed over to a public consultation process, and generation of its legitimacy in the course of the process. Ultimately, however, the role of scientists in this second dimension is rather limited, notwithstanding the explicit request from government ministries for scientific advice on the present case of the Regional Climate Change Adaptation Plan for the MRS.

Ad (3): As mentioned earlier, successful implementation of the first dimension as well as appropriate communication between science and policy/practice are two preconditions for final implementation of the measures as referred to in the third dimension. It cannot, however, be assumed that a successful process will immediately lead to concrete adaptation action in terms of this last dimension. Here, competences, capacities and the accountability of local actors, as well as linkages to national frameworks, regional guidance and local action are vital (e.g., Mees et al. 2012; Satterthwaite et al. 2007; Bulkeley et al. 2009). This is where the third dimension comes to the fore: the execution of concrete adaptation measures that confront climate change impacts with "real" action.

The current chapter focuses on the third dimension. It analyses the potential of realizing measures such as cool roofs for indoor cooling (cf. Chap. 7) or new irrigation technologies to save water (cf. Chap. 4) at the municipal or even individual level, once the second implementation dimension has been successfully completed. Hence the various factors that constitute a challenge to progress of the measures in the MRS at this stage are discussed, based on a literature review and the perspective and understanding of the stakeholders concerned.

So far, little is known about the success or failure of implementing such measures (e.g., Amundsen et al. 2010; Bulkeley et al. 2009). Most articles on the implementation of adaptation measures are conceptual and tend to adopt a theoretical-economic perspective (see e.g., Berkhout 2005; Mendelsohn 2006; IPCC 2007; Osberghaus et al. 2010). Financial limitations are often considered major constraints on the implementation of climate change measures (e.g., Bulkeley et al. 2011; Amundsen et al. 2010; Lidskog and Elander 2010; Lorenzoni et al. 2000). Mees et al. (2012) point to the empirical exploration of governance arrangements in the context of implementing climate adaptation, taking green-roof policies as a concrete example. They conclude among other things that a predominantly public responsibility is both feasible and indispensable to the introduction and successful functioning of green roofs.

Based on the experience of developing a Regional Climate Change Adaptation Plan for Santiago de Chile, the hypothesis behind the argumentation in this chapter is that (1) political will, (2) the existing knowledge and information base, (3) institutional arrangements and (4) financial means are four crucial elements when it comes to implementing adaptation measures. Challenges that may arise in the course of implementing concrete adaptation measures will therefore be debated in the following sections.

Section 10.2 discusses specific preconditions for the implementation of adaptation measures in the MRS and, in light of the actors involved, presents some online survey results containing key information for the assessment of competences and capacities in the context of implementation, while Sect. 10.3 analyses challenges and constraints. Section 10.4 focuses on the capacities and competences of the actors involved in implementing the measures. Section 10.5 concludes with some remarks on lessons learned in the course of overcoming implementation constraints in the MRS.

10.2 Specific Preconditions for the Implementation of Adaptation Measures in the MRS

The process of developing adaptation measures for the MRS was accompanied by cross-cutting action between sectors and administrative levels in an attempt to address the complexity of climate change and the interwoven processes it involves, and likewise to strengthen the exchange and cooperation between them. This ties in with the assumption that a variety of factors are frequently related to local administrative structures, for example, or available means and can either hinder or benefit the implementation of adaptation measures. Accordingly, this section assesses some characteristics of the MRS in the context of implementing climate change response in the form of adaptation measures.

The argumentation is based on the results of an online survey conducted at the end of the participatory process prior to handing over the Regional Climate Change Adaptation Plan for the MRS to the authorities responsible for its implementation. The survey was online for three months and stakeholders who participated at least once in the process over the two-and-a-half-year period were contacted via email with a request for feedback on

1. the transdisciplinary approach that formed the basis of the overall process regarding
   (a) the participatory process in the form of roundtable meetings and
   (b) knowledge transfer from science to praxis;
2. information on the implementation of the measures and willingness to participate in follow-up activities.

The questionnaire was sent to the 69 people who took part in at least one of the nine roundtable meetings held before the survey went online. It yielded a total of 23 responses (33 %), 15 of which were completed questionnaires and eight partly completed. Since the questions called for individual reflection, the eight incomplete questionnaires were also taken into consideration in order to gain a more comprehensive picture. Several follow-ups in the form of reminders were sent out via email and, in addition, some personal communication took place in an attempt to achieve a satisfactory response rate. It is nonetheless possible that stakeholders who attended only a few roundtable meetings were less committed to or engaged in the process and less willing to respond to the survey, leading to a decrease in the overall response rate.

On the whole, a response rate of 33 % can be regarded as satisfactory. Response rates of online surveys are highly divergent, ranging from five to over 50 % in some cases (cf. Nulty 2008; Couper 2000; Archer 2008). Response rates depend on a range of factors, the most prominent of which have been identified as the topic of the study and its salience, the length of the questionnaire, the incentives offered, prenotifications and reminders sent, and personalized notifications (cf. Dillman 2007; Deutskens et al. 2004; Monroe and Adams 2012; Archer 2008).

The uneven participation by stakeholders was problematic when it came to interpreting answers to the questionnaire. Bearing in mind the qualitative character of this evaluation study, however, the results give a good impression of the stakeholder perspective and are considered a basis for further assumptions and analyses.

The results of the online survey on the participatory process reveal that the joint roundtable meetings over the two-and-a-half-year time period helped participating actors and sectors to establish contact, to build relations of trust, and to set up networks. The stakeholders in the survey asserted that their expectations had been fulfilled, especially with regard to their knowledge growth on climate change and adaptation. It can therefore be concluded that reinforcing knowledge and adding new information pertaining to the specific needs of the MRS enabled the actors involved to expand their adaptive capacity during the process.[1]

Adaptive capacity is generally seen as a "positive" property. This section, however, also considers constraints identified by participants as affecting the process of developing adaptation measures for the MRS. As far as the principal outcome of the process is concerned, a majority of respondents (73 %) sees the proposed measures as covering the most pressing needs of the MRS in the face of climate change impacts. But, a possible constraint is the priority given to developing climate change adaptation measures over other public concerns. Whereas climate change and its impacts call for a long-term perspective, political agendas have a more short-term orientation, a conceivable reason for lack of awareness about today's climate change impacts and workable interventions. Furthermore, although future climate change projections remain uncertain, action should take place under these conditions. Bearing this in mind, current and possible future impacts of climate change were worked on and short- to long-term measures for the Santiago case proposed in an effort to underpin the potentially positive effect of direct interventions. Since climate change was confirmed as a relevant topic for Santiago by almost all respondents following the participatory process, it will now hopefully achieve greater attention on the political agenda and become a starting point for action.

The intention in the following section is to link these "preconditions", identified as the overall ("positive" and "negative") outcome of the process, to the general capacity of the various actors to take action, and to discuss their roles and responsibilities in the context of implementing adaptation measures. The objective here is to argue on the basis of feedback from the actors involved in the process rather than to undertake an actor or/and governance analysis.

---

[1] Adaptive capacity is seen here as the ability to adapt, while IPCC (2007) defines it as an "adjustment in natural or human systems in response to actual or expected climatic stimuli or their effects". Adger et al. (2005) state that "adaptation can involve both building adaptive capacity, thereby increasing the ability of individuals, groups or organisations to adapt to changes, and implementing adaptation decisions, i.e., transforming that capacity into action". Arguing according to Engle (2010), Frommer (2009), Smit and Pilifosova (2001) among others, increased knowledge and the establishment of contact and trust are preconditions for successful implementation of the measures concerned. And even more concrete in the words of Adger et al. (2005), "actions associated with building adaptive capacity may include communicating climate change information, building awareness of potential impacts, maintaining well-being, protecting property or land, maintaining economic growth, or exploiting new opportunities". To build the adaptive capacity of local authorities, such aspects as civil society values and knowledge, and the data and results from the broader analytical community, e.g., research, consultancy and sector authorities, need to be considered (e.g., Füssel and Klein 2006; Agrawala and van Aalst 2008). This implies that interaction between values, science and policy exists (Betsill and Bulkeley 2006) or can be established.

10.3 Challenges and Constraints: What Hinders or Promotes the Implementation of Adaptation Measures in the MRS?

In order to figure out the specific constraints of implementing the 14 climate change adaptation measures elaborated in the participatory process and contained in the Regional Climate Change Adaptation Plan for the MRS, the results of the online survey are clustered along four determining factors: (1) political will, (2) knowledge and information base, (3) institutional arrangements, and (4) financial means.

In the following, a brief review of the four determining factors as seen by other scholars is presented before the survey results are discussed.

1. *Political will* and an interest in action seems to be the prerequisite for placing, maintaining and pushing the topic on the political agenda (cf., e.g., Eisenack and Stecker 2012; Preston et al. 2011; Heinrichs et al. 2011; Biesbroek et al. 2010; Sánchez-Rodríguez et al. 2008; Füssel 2007).
2. The complexity of the issues associated with climate change, impact and adaptation calls for a reliable *knowledge and information base*. This implies interdisciplinary work and should also consider information on how to deal, for example, with uncertainties and the characteristics of local decision-making (Krellenberg and Barth 2014; Fankhauser and Tol 1997; Smit and Pilifosova 2001; Ahmed et al. 1999).
3. *Institutional arrangements* play a decisive role since administrative structures are usually path-dependent and somewhat inflexible on the question of integrating complex new topics. The capacity to coordinate several levels of decision-making and several policy fields is paramount to the successful implementation of adaption measures (e.g., Bauer et al. 2011; Amundsen et al. 2010; Burch and Robinson 2007, 2011; Corfee-Morlot et al. 2009; Bulkeley et al. 2011; Laukkonnen et al. 2009; Sánchez-Rodríguez et al. 2008).
4. Having the *financial means* to support the implementation of specific measures is indispensable. Apart from technical infrastructure this also means investment in manpower and capacity building (cf., e.g., Runhaar et al. 2012; Bulkeley et al. 2011; Heinrichs et al. 2011; Hunt and Watkiss 2011; Measham et al. 2011; Preston et al. 2011; Amundsen et al. 2010; Corfee-Morlot et al. 2009).

Concerning concrete conditions in the MRS, the following observations were made:

Ad 1: With regard to the first aspect, the importance of political will, lack of willingness on the part of decision-makers to put the topic on the political agenda and push it further was identified by the polled participants (69 %) as the chief obstacle to the implementation of adaptation measures. Some respondents referred to the poor political commitment of national actors during the two-and-a-half-year process to implement specific measures. The advisability, for instance, of having expected the participation of more senior decision-makers in the participatory process, particularly from the national level, was also mentioned as a limiting factor, since Chile's centralism sees the national level playing a major role in the effective and sustainable implementation of the measures. 76 and 64 % respectively of survey respondents saw the ministries and the national government as key actors for implementation. This "implementation activity", however, relates to the second rather than the third implementation dimension (cf. Sect. 10.1), where concrete measures are drawn up.

Ad 2: Regarding the second aspect, the existence of knowledge and information, a number of participants (44 %) identified poor information on the implementation process as a limiting factor. Consequently, 87 % of respondents stressed the benefit of an implementation guideline. It is debatable whether they argued for a guideline knowing it would be one of the overall products accompanying the Plan or merely expressed a personal wish. In the event, the Manual for Implementation was provided at the end of the process but was not yet available when the survey went online. Respondents viewed the guideline primarily as a method of presenting the responsibilities and procedures involved, as information for those who had not participated in the process (information diffusion), as practical details to facilitate implementation, and as support material to achieve objectives. With its wealth of information, the Manual of Implementation complements the Regional Climate Change Adaptation Plan.

Ad 3: Referring to the third aspect, institutional arrangements, survey respondents addressed the decisive role of institutional coordination. They claimed coordination of administration departments was highly relevant to implementation, while more than half of the polled participants categorized collaboration between public and private actors as crucial to the process. Furthermore, (active or passive) involvement of the population was identified as relevant and since public opinion of the measures was seen as influential, 44 % of respondents saw lack of public acceptance as a major barrier to implementation. This highlights the importance of participation in climate change decision-making, backed by a range of scholars (e.g., Anguelovski and Carmin 2011; Lidskog and Elander 2010; Puppim de Oliveira 2009), especially in the context of climate change adaptation (Huntjens et al. 2012; Jones et al. 2012; Termeer et al. 2012).

Yet another governance and institutional element emphasized in the context of implementation was the somewhat foggy distribution of roles and functions, which could be due to the strong sector divide that characterizes the Chilean administration. This need for clear definition and unambiguous distribution of roles and responsibilities has been singled out by others, who argue that it is fundamental to fostering adaptation action, regardless of the complex actor relationships and governance challenges entailed in managing adaptation (cf. e.g., Termeer et al. 2012; Mees et al. 2012; Corfee-Morlot et al. 2009). Here, respondents labelled the coherence of adaptation measures with national objectives as relevant to their achievement. This underpins the significance of relying on the support and the interest of the national level and relates to the fact that the central Chilean government has a strong influence on urban issues in general (cf. Hölzl et al. 2012), whereas competencies of the Metropolitan Region are ill-defined, resources to act autonomously inadequate, and their capacity for action somewhat restricted (Chuaqui and Valdevieso 2004).

Although on the whole Chilean municipalities have a weak position compared to higher decision-making levels, and their power and intervention capacity depends on poor resources (cf. Hölzl et al. 2012), they nevertheless display the potential to act independently. One option is to organize their own budgets or apply for additional funds, once a regional framework such as the Regional Climate Change Adaptation Plan has come into force. However, assuming that legal modifications and additional funding will be required to realize adaptation measures at regional and local level, the power of the central level is ratified. Thus, for successful implementation of the measures at regional and municipal level, the consolidation of adaptation on the national agenda should likewise be achieved. (cf. García Soler 2013).

Ad 4: With regard to the fourth factor, financial means, insufficient funding for the measures is seen by the polled participants as one of the main barriers to implementation. Fifty per cent consider financial means essential to implementation. This is consistent with Obreque (2011), who concludes that the availability of economic means is one of the most significant factors determining concrete implementation of the measures included in the National Action Plan for Climate Change (PANCC). Lack of financial means and consequently the obstacle to implementing measures could be even more acute in the case of the municipalities, which often depend heavily on subsidies from higher decision-making levels. The economic capacity of the municipalities in the MRS, however, is highly diverse (SINIM n.d.). The more "wealthy" may be in a position to realize some of the adaptation measures, especially if legal or administrative rearrangements do not require national approval.

10.4 Capacities and Competences: The Actor's Role in Implementing Adaptation Measures

The factors, constraints and challenges associated with implementing adaptation measures in the MRS identified by the polled participants underpin the key role that the capacities and competences of the actors and sectors involved play in this context.

Given the cross-cutting character of climate change adaptation in terms of action fields and decision-making levels, the institutional capacity to incorporate and integrate adaptation measures in existing policies, plans and programmes is vital to implementation (cf. Bulkeley et al. 2009, 2011; Hunt and Watkiss 2011; Sánchez-Rodríguez et al. 2008). In other words, the actors concerned are required to have specific capacities and competences that go beyond traditional expertise (i.e., sector-specific knowledge; dealing with well-known problems; routine-oriented action). Fostering institutional capacity at regional and municipal level calls for a multi-level approach that enables all these actors to play a part in climate change response (cf. Amundsen et al. 2010; Betsill and Bulkeley 2006; Cash et al. 2006). Birkmann et al. (2010) also speak of an integrative approach, arguing that new "adaptive urban governance" forms are needed if all issues affected are to be covered and the relevant agents involved. Coordination, mutual reinforcement policies across sectors, innovation, transparency, accountability, flexibility, involvement of stakeholders and population, and the establishment of linkages with other matters are some of the elements that come into play in such new governance arrangements (cf. Anguelovski and Carmin 2011; Tanner et al. 2009).

As implementation finally takes place at the municipal level, the extent to which this is possible will depend on local level capacities, on the one hand, and multi-level governance arrangements, on the other (Satterthwaite et al. 2007; Amundsen et al. 2010). According to Corfee-Morlot et al. (2009) participation, strategic planning, provision of an analytical basis for short- and long-run planning, cost efficiency, encouragement of experimentation and innovation, monitoring and evaluation are among the factors that could enhance the capacities of the actors involved. Huntjens et al. (2012) add flexibility, reflection, learning and innovation, trust building, collective choice arrangements, policy experimentation, and conflict resolution mechanisms. Whatever this list of requirements looks like: it is obvious that traditional and well-established forms of policy and administration must be overcome.

Returning to the "real case" of the MRS, the need to involve the municipalities in the process of implementation was underlined by all survey respondents, with reference to their importance as managers of basic territorial units and their direct contact to the population. An additional workshop organized with representatives of three selected municipalities of the MRS helped to identify specific challenges and needs at municipal level associated with implementing adaptation measures.

In some cases these characteristics are linked to personality—innovative policies are frequently designed and put into effect by charismatic and powerful "leaders". In other cases learning is a decisive step in developing these competences and capacities. In the MRS, the cycle of roundtable meetings provided a platform for capacity building by fostering knowledge exchange, a podium to learn from each other and establish contact face to face in continuous meetings. Although maintaining participative processes during implementation of the measures would be desirable, continuity is now in the hands of the participants and political entities, since the academic contribution to the adaptation process came to an end when the Plan was officially handed over.

10.5 Concluding Remarks

As the case of Santiago de Chile substantiates, implementation of adaptation measures is by no means an easy task. Several determining factors identified in the literature and supplemented by the statements of the stakeholders involved in the development of the Regional Climate Change Adaptation Plan for the MRS coincide with those of other case studies.

- **Leadership and political** will are indispensable to legitimizing the final implementation of climate change response. In this sense it should once again be highlighted that the overall process undertaken in the Metropolitan Region of

Santiago de Chile proved successful only as a result of the commitment of key personalities. It is likewise assumed that final implementation of the measures will depend on the personal engagement and commitment of the actors at municipal, regional and national level who are willing to take on the responsibility.

However, leadership and political will is not everything. **Information** on what is to be implemented and how also plays a major role. Here it was almost certainly beneficial in terms of the final steps to be undertaken that regional climate change, climate change impacts and implementation issues were discussed in-depth with key partners during the entire process (see Chap. 9). It was vital that final implementation did not present itself as a surprise element at the end.

- Furthermore, a number of **constraints** that emerged during the first implementation dimension (i.e., implementation of the transdisciplinary approach) were overcome by direct interventions before they became "barriers". It is assumed that the possibility to intervene will prevent the occurrence of major problems during the third implementation phase. In addition, the attempt to link the 14 adaptation measures to existing policies and programmes could contribute positively to their implementation. The feasibility of implementing the measures with regard to institutional and legal frameworks and co-benefits in the frame of scoping studies will hopefully also benefit implementation. There may, however, be some insurmountable barriers attached to the legal framework.
- The **transdisciplinary approach** was a concerted effort to open the floor to successful implementation of the adaptation measures. An overwhelming majority of online survey respondents (92 %) called for roundtable meetings during the implementation and monitoring phases. Continuity of the participatory process could help to coordinate the various actors involved and overcome the existing sector divide. Similarly, evaluation of the evolution of the measures could be steered at the science-policy interface.
- Coordination and contribution of **international experts** was seen as positive, since experiences in other countries could be of relevance when it comes to launching the adaptation measures. In this context, the polled participants were asked which actors should lead the process to expand and develop the Regional Climate Chance Adaptation Plan in the future when the international cooperation process had come to an end. Interestingly, national level actors were named for the most part, whereas few referred to regional entities (SEREMI MA or/and GORE).

In this regard, some lessons can already be learned from the Santiago de Chile case but as final implementation is still pending it is important to keep on exchanging knowledge between cities and countries in order to avoid maladaptation and to try to speed up the processes of implementation.

References

Adger, W. N., Arnell, N. W., & Tompkins, E. L. (2005). Successful adaptation to climate change across scales. *Global Environmental Change, 15*, 77–86.

Agrawala, S., & van Aalst, M. (2008). Adapting development cooperation to adapt to climate change. *Climate Policy, 8*(2), 183–193.

Ahmed, A. U., Alam, M., & Rahman, A. A. (1999). Adaptation to climate change in Bangladesh: Future outlook. In S. Huq, M. Asaduzzaman, Z. Karim, & F. Mahtab (Eds.), *Vulnerability and adaptation to climate change for Bangladesh* (pp. 125–143). Dordrecht: J. Kluwer Academic Publishers.

Amundsen, H., Berglund, F., & Westskog, H. (2010). Overcoming barriers to climate change adaptation – a question of multilevel governance? *Environment and Planning C: Government and Policy, 28*, 276–289.

Anguelovski, I., & Carmin, J. (2011). Something borrowed, everything new: innovation and institutionalization in urban climate governance. *Current Opinion in Environmental Sustainability, 3*, 169–175.

Archer, T. M. (2008). Response rates to expect from web-based surveys and what to do about it. Research in brief. *Journal of Extension*. http://www.joe.org/joe/2008june/rb3.php. Accessed 22 Mar 2013.

Bauer, A., Feichtinger, J., & Steurer, R. (2011). *The governance of climate change adaptation in ten OECD countries: Challenges and approaches*. Discussion Paper 1-2011. In FER (Institute of Forest, Environmental, and Natural Resources Policy). http://www.wiso.boku.ac.at/fileadmin/_/H73/H732/_TEMP_/InFER_DP_11_1__The_Governance_01.pdf. Accessed 15 Aug 2013.

Berkhout, F. (2005). Rationales for adaptation in EU climate change policies. *Climate Policy, 5*(3), 377–391.

Betsill, M., & Bulkeley, H. (2006). Cities and the multilevel governance of global climate change. *Global Governance, 12*, 141–159.

Biesbroek, G. R., Swart, R. J., Carter, T. R., Cowan, C., Henrichs, T., Mela, H., et al. (2010). Europe adapts to climate change: Comparing national adaptation strategies. *Global Environmental Change, 20*, 440–450.

Birkmann, J., Garschagen, M., Kraas, F., & Quang, N. (2010). Adaptive urban governance: New challenges for the second generation of urban adaptation strategies to climate change. *Sustainability Science, 5*, 185–206.

Bulkeley, H., Schroeder, H., Janda, K., Zhao, J., Armstrong, A., Chu, S. Y., et al. (2009). *The role of institutions, governance, and urban planning for mitigation and adaptation*. Report for the World Bank urban symposium on climate change. http://www.eci.ox.ac.uk/publications/downloads/bulkeley-schroeder-janda09.pdf. Accessed 15 Aug 2013.

Bulkeley, H., Schroeder, H., Janda, K., Zhao, J. Armstrong, A., Yi Chu, S., et al. (2011). The role of institutions, governance, and urban planning for mitigation and adaptation. In D. Hoornweg, M. Freire, M. J. Lee, P. Bhada-Tata, & B. Yuan (Eds.), *Cities and climate change* (Responding to an urgent agenda, pp. 125–159). Washington, DC: The World Bank.

Burch, S., & Robinson, J. (2007). A framework for explaining the links between capacity and action in response to global climate change. *Climate Policy, 7*(4), 304–316. http://dx.doi.org/10.1080/14693062.2007.9685658.

Cash, D. W., Cash, W. N., Adger, F., Berkes, P., Garden, L., Lebel, P., et al. (2006). Scale and cross-scale dynamics: Governance and information in a multilevel world. *Ecology and Society, 11*(2), 8.

Chuaqui, T., & Valdevieso, P. (2004). Una ciudad en busca de un gobierno: una propuesta para Santiago. *Revista de Ciencia Politica, 24*(1), 104–127.

Corfee-Morlot, J., Kamal-Chaoui, L., Donovan, M. G., Cochran, I., Robert, A. & Teasdale, P.J. (2009). *Cities, climate change and multilevel governance*. OECD environmental working papers, number 14. http://www.oecd.org/gov/regionaldevelopment/44232263.pdf. Accessed 15 Aug 2013.

Couper, M. P. (2000). Web surveys. A review of issues and approaches. *Public Opinion Quarterly, 64*, 464–494.

Deutskens, E., De Ruyter, K., Wetzels, M., & Oosterveld, P. (2004). Response rate and response quality of internet-based surveys: An experimental study. *Marketing Letters, 15*(1), 21–36.

Dillman, D. A. (2007). *Mail and internet surveys: The tailored design method*. Hoboken, NJ: John Wiley & Sons.

Eisenack, K., & Stecker, R. (2012). A framework for analyzing climate change adaptations as actions. *Mitigation and Adaptation Strategies for Global Change, 17*(3), 243–260.

Engle, N. L., & Lemos, M. C. (2010). Unpacking governance: Building adaptive capacity to climate change of river basins in Brazil. *Global Environmental Change, 20*(1), 4–13.

Fankhauser, S., & Tol, R. S. J. (1997). The social costs of climate change: The IPCC second assessment report and beyond. *Mitigation and Adaptation Strategies for Global Change, 1*, 385–403.

Frommer, B. (2009). Handlungs- und Steuerungsfähigkeit von Städten und Regionen im Klimawandel. *Raumforschung und Raumordnung, 67*(2), 128–141.

Füssel, H. M. (2007). Adaptation planning for climate change: Concepts, assessment approaches, and key lessons. *Sustainability Science, 2*(2), 265–275.

Füssel, H. M., & Klein, J. T. (2006). Climate change vulnerability assessments: An evolution of conceptual thinking. *Climatic Change, 75*, 301–329.

García Soler, N. (2013). *The cross-cutting character of adaptation to climate change*. Motivations and obstacles in the case of the metropolitan region of Santiago de Chile. Unpublished master's thesis. University of Leipzig. Leipzig.

Heinrichs, D., Aggarwal, R., Barton, J., Bharucha, E., Butsch, C., Fragkias, M., et al. (2011). Adapting cities to climate change: Opportunities and constraints. In D. Hoornweg, M. Freire, M. J. Lee, P. Bhada-Tata, & B. Yuan (Eds.), *Cities and climate change* (Responding to an urgent agenda, pp. 193–224). Washington, DC: The World Bank.

Heinrichs, D., Krellenberg, K., & Fragkias, M. (2013). Symposium urban responses to climate change: Theories and governance practice in cities of the global south. *International Journal of Urban and Regional Studies, 37*(6), 1865–1878.

Hölzl, C., Nuissl, H., Höhnke, C., Lukas, M., & Rodriguez, C. (2012). Dealing with risks: A governance perspective on Santiago de Chile. In D. Heinrichs, K. Krellenberg, B. Hansjürgens, & F. Martinez (Eds.), *Risk habitat megacity* (The case of Santiago de Chile, pp. 87–108). Heidelberg: Spinger.

Hunt A., & Watkiss P. (2011). Climate change impacts and adaptation in cities: A review of the literature. *Climatic Change, 104*(1), 13–49 (Special Issue: Understanding climate change impacts. *Vulnerability and Adaptation at City Scale*). doi: 10.1007/s10584-010-9975-6.

Huntjens, P., Lebel, L., Pahl-Wostl, C., Campkin, J., Schulze, R., & Kranz, N. (2012). Institutional design propositions for the governance of adaptation to climate change in the water sector. *Global Environmental Change, 22*, 67–81.

Huq, S., Reid, H., & Murray, L. A. (2006). Gatekeeper (123). *Climate Change and Development Links*, May 2006 – IIED. http://pubs.iied.org/pdfs/14516IIED.pdf. Accessed 15 Aug 2013.

IPCC. (2007). *Climate change 2007: Impacts, adaptation and vulnerability*. Contribution of working group II to the fourth assessment report of the IPCC. http://www.ipcc.ch/pdf/assessment-report/ar4/wg2/ar4_wg2_full_report.pdf. Accessed 15 Aug 2013.

Jones, N., Clark, J., & Tripidaki, G. (2012). Social risk assessment and social capital: A significant parameter for the formation of climate change policies. *The Social Science Journal, 49*, 33–41.

Krellenberg, K., & Barth, K. (2014). Planning collaborative climate change adaptation responses in the Metropolitan Region of Santiago de Chile. *Interdisciplinary Science Reviews, 39*(4).

Krellenberg, K., Jordán, R., Rehner, J., Schwarz, A., Infante, B., Barth, K., et al. (2013). *Adaptación al cambio climático en megaciudades de América Latina. Red Regional de Aprendizaje del Proyecto ClimaAdaptaciónSantiago* (CAS). Comisión Económica para América Latina y el Caribe (CEPAL).

Kumar, C. B. (2013). Climate change and Asian cities: So near yet so far. *Urban Studies, 50*, 1456.

Laukkonen, J., Blanco, P. K., Lenhart, J., Kleiner, M., Cavric, B., & Kinuthia-Njenga, C. (2009). Combining climate change adaptation and mitigation measures at the local level. *Habitat International, 33*, 287–292.

Lidskog, R., & Elander, I. (2010). Addressing climate change democratically. Multi-level governance, transitional networks and governmental structures. *Sustainable Development, 18*, 32–41.

Lorenzoni, I., Jordan, A., Hulme, M., Turner, R. K., & O'Riordan, T. (2000). A co-evolutionary approach to climate change impact assessment (I): Integrating socioeconomic and climate change scenarios. *Journal of Global Environmental Change, 10*(1), 57–68.

Measham, T. G., Preston, B. L., Smith, T. F., Brooke, C., Gorddard, R., Withycombe, G., et al. (2011). Adapting to climate change through local municipal planning: Barriers and challenges. *Mitigation Adaptation Strategies Global Change, 16*(8), 889–909.

Mees, H. L. P., Driessen, P. P. J., & Runhaar, H. A. C. (2012). Exploring the scope of public and private responsibilities for climate adaptation. *Journal of Environmental Policy and Planning, 14*(3), 305–330.

Mendelsohn, R. (2006). The impacts of climate change and its policy implications. *Ambio, 35*(6), 273.

Monroe, M. C., & Adams, D.C. (2012). Increasing response rates to web-based surveys. *Journal of Extension, 50*(6), 6TOT7. http://www.joe.org/joe/2012december/tt7.php. Accessed 15 Aug 2013.

Nulty, D. D. (2008). The adequacy of response rates to online and paper surveys: What can be done? *Assessment and Evaluation in Higher Education, 33*(3), 301–304.

Obreque, F. (2011). *Informe final*. Evaluación de Medio Término del Plan de Acción Nacional de Cambio Climático. http://www.sinia.cl/1292/w3-article-51202.html. Accessed 15 Aug 2013.

Osberghaus, D., Finkel, E., & Pohl, M. (2010). *Individual adaptation to climate change: the role of information and perceived risk*. ZEW Discussion Paper No. 10-061. http://mpra.ub.uni-muenchen.de/26569/1/MPRA_paper_26569.pdf. Accessed 15 Aug 2013.

Preston, B. L., Westaway, R. M., & Yuen, E. J. (2011). Climate adaptation planning in practice: An evaluation of adaptation plans from three developed nations. *Mitigation, Adaptation, Strategies Global Change, 16*, 407–438.

Puppim de Oliveira, J. A. (2009). The implementation of climate change related policies at subnational level: An analysis of three countries. *Habitat International, 33*, 253–259.

Runhaar, H., Mees, H., Wardekker, A., Van der Sluijs, J., & Driessen, P. P. J. (2012). Adaptation to climate change-related risks in Dutch Urban Areas: Stimuli and barriers. *Regional Environmental Change, 12*(4), 777–790.

Sánchez-Rodríguez, R., Fragkias, M. and Solecki, W. (2008). *Urban responses to climate change a focus on the Americas*. Workshop Report. June 2008. http://ccsl.iccip.net/ur2cc.pdf. Accessed 15 Aug 2013.

Satterthwaite, D., Huq, S., Reid, H., Pelling, M., & Romero, P. (2007). *Adapting to climate change* in urban areas: the possibilities and constraints in low- and middle- income nations. Paper series: climate change cities – 1. London: IIED Human Settlements Discussion.

SINIM (National System of Municipal Information). n.d. Sistema Nacional de Información Municipal. http://www.sinim.cl/. Accessed 20 Mar 2013.

Smit, B. & Pilifosova, O. (2001). Adaptation to climate change in the context of sustainable development and equity. In J.J. McCarthy (Ed.). *Climate change 2001. Impacts, adaptation, and vulnerability; Contribution of working group II to the third assessment report of the intergovernmental panel on climate change* (1st ed., pp. 877–912). Cambridge: Cambridge University Press.

Tanner, T., Mitchell, T., Polack, E., & Guenther, B., (2009). *Urban governance for adaptation: assessing climate change resilience in Ten Asian Cities*. Working paper 315, Brighton, UK: Institute of Development Studies (IDS). http://www.ids.ac.uk/files/dmfile/Wp315.pdf. Accessed 15 Aug 2013.

Termeer, C. J. A. M., van Vliet, M., Berkhout, F. G. H., Driessen, P., Leroy, P., van Rijkswick, H. F. N. W., et al. (2012). *The governance of adaptation to climate change. A collaborative action research program to develop and test legitimate, effective and resilient governance arrangements for climate adaptation*. Midterm review. KfC report number 63/2012.

Going Beyond Santiago: A Regional Learning Network for Climate Change Adaptation in Latin American Megacities

11

Ricardo Jordan, Kerstin Krellenberg, and Johannes Rehner

Abstract

Adaptation to climate change is a highly context-specific field of political action. Not only do the features and impacts of local climate change vary from one area to another, political, economic and social contexts also generate diversity in terms of urgency and priority of concrete adaptation measures. In Latin America, metropolitan regions are the focus of demographic change and economic activity, concentrating high levels of poverty and consequently vulnerability to the risks produced by climate change. Hence, the need for specific climate change adaptation plans, policies and measures was discussed in a Regional Learning Network initiated within the context of an international research project.

This chapter summarizes the results of three workshops designed to address the expected features and impacts of climate change in six Latin American megacities and, focusing on adaptation, to examine how climate change policy is currently defined and implemented. In conclusion, it discusses concrete measures. In all of the cities under review, steps have been taken to institutionalize climate change response on an urban scale—albeit the approaches adopted and experiences gained are quite different. The measures in question prove more successful if linked to strategic goals or national urban policies, a shrewd move that produces co-benefits. The potential of these elements of climate change adaptation to pave the way for more sustainable cities without the obvious link to climate change indicates that ascertaining the exact dimension of climate change in the urban centres is not a high priority.

R. Jordan (✉)
Division of Sustainable Development and Human Settlements, UN ECLAC, Santiago, Chile
e-mail: ricardo.jordan@cepal.org

K. Krellenberg
Department of Urban and Environmental Sociology, Helmholtz Centre for Environmental Research – UFZ, Permoser Str. 15, 04318 Leipzig, Germany

J. Rehner
Pontificia Universidad Católica de Chile, Santiago, Chile
Consultant UN ECLAC, Santiago, Chile

K. Krellenberg and B. Hansjürgens (eds.), *Climate Adaptation Santiago*,
DOI 10.1007/978-3-642-39103-3_11, © Springer-Verlag Berlin Heidelberg 2014

Keywords
Climate change adaptation • Latin America • Megacities

11.1 Latin American Cities and Climate Change

Latin America and the Caribbean is the most urbanized region in the developing world. According to censuses conducted in 2010, urbanization is an ongoing process in which the urban population has soared to an approximate 80 % (UN ECLAC 2012). Cities—especially large cities—are simultaneously culprits and victims of climate change (see Chap. 1 of this book), making the comparison of megacities in Latin America a stimulating research topic in terms of expected climate change, its associated impacts and ongoing and/or planned adaptive response (c.f. Jordán 2008).

This chapter presents research findings linked to these issues from the perspective of six selected megacities in the Latin American Region: Buenos Aires (Argentina), Bogotá (Columbia), Lima (Peru), Mexico (Mexico), Sao Paulo (Brazil) and Santiago de Chile (Chile). The selection builds on previous research carried out on the sustainability of these cities (Jordán et al. 2010). The findings show that all six have a number of aspects in common:

- They show evidence of a steady population growth, income enhancement and increased economic wealth. This in turn creates serious pressure on natural resources rooted in the growing consumption-based demand for water, food and energy.
- They display an ongoing pattern of socio-spatial segregation and higher density combined with suburban growth.
- They possess complex governance structures that can amplify or even produce risks to and negative impacts on human security.

Large cities are the focus of economic development. In the time scale most relevant to a discussion on climate change (here 2050), the global urban population is projected to reach over six billion inhabitants or two-thirds of the world population (UN DESA 2008). Regardless of whether this trend has its origins in general demographic changes and migration patterns (UN ECLAC 2012), in the persistent agglomeration of economic activities in space (Krugman 1997; The World Bank 2009) or in the nature of the capitalist accumulation process itself, metropolitan areas are and will remain the focus of economic activities and social challenges. This is especially true for large Latin American cities.

So far Latin American cities have confined climate change response action primarily to greenhouse gas emission reduction, e.g., via Clean Development Mechanism (CDM) activities. They also focus on the climate change-related risks already in evidence, such as floods, heat stress, droughts and landslides (Satterthwaite 2006; IIED 2007; Romero Lankao 2010; Krellenberg 2012; Heinrichs et al. 2011). It is important in this context to highlight the per capita carbon emissions of cities in the "global south", notably Latin America, which are relatively low (ONU Habitat 2012). Hence the debate on the linkage between urbanization trends and carbon emissions is mainly driven by the experience of Europe and North America. In Latin America, climate change is seen as an environmental problem first and foremost related to development issues, with a major impact on the population that calls for adequate response action at the urban level. According to Romero Lankao (2007) this action is or should differ from responses embarked upon in the "global north". In other words, despite technology, experience and best practice, which may well be of universal significance, the respective regional and local solutions remain context-specific.

To find out how climate change is manifested in the six selected cities and how response is organized, the findings presented in this chapter present a summary of the documentation of the three workshops undertaken within the framework of an established Regional Learning Network (RLN) (Krellenberg et al. 2013a). They build on presentations and discussions with scientists and representatives from government institutions in the cities in question and are further enriched by a review of the literature. The Network sought to bring a variety of actors and cities together in order to learn from the experience of each city in adapting to climate change and to discuss transferable solutions.

Building on the results of the Regional Learning Network, the chapter is structured as follows: Sect. 11.2 synthesizes observed changes in local climates, climate change projection data on the metropolitan scale, downscaling techniques applied to each city, and the expected impacts of climate change. Based on this information, Sect. 11.3 outlines the need for climate change in the six selected cities. Section 11.4 summarizes their existing climate change strategies and plans, highlighting the chief content and the challenges the implementation experience entails. Section 11.5 discusses key findings of the analysis of the different paths taken in each city. The chapter winds up with some concluding remarks on lessons learned from the case cities and on transferable solutions (Sect. 11.6).

11.2 Climate Change and Its Impacts

Debating urban adaptation plans, strategies and measures calls for in-depth information on projected climate changes at urban-regional level based on current and historic climate data. Data analysis brings with it per se a marked uncertainty that should be carefully weighed up, especially when it comes to data comparison. Bearing this in mind, this section summarizes existing knowledge on the prevailing climate in the six selected cities. It refers firstly to the individual characteristics of the climate conditions in each city (Table 11.1) and, secondly, recapitulates the expected trends in climate change (Table 11.2).

As indicated in Table 11.1, the six cities under review represent a broad spectrum of climatic conditions that demand consideration in the discussion on climate change policies. One is located in a very dry climate (Lima), whereas Santiago de Chile, although endowed with very dry summers, has winter rainfalls. Located on the other side of the Andeans, Sao Paulo and Buenos Aires both have a

Table 11.1 Climate features of the six Latin American megacities (*Source*: Krellenberg et al. (2013a: 34), based on RLN workshops; data on precepitation and temperature refer to observed long-term averages taken from different national meteorological services: Bogotá http://www.ideam.gov.co; Buenos Aires http://www.smn.gov.ar; Lima INEI (2007: 101); Mexico City http://www.smn.cna.gob.mx; Santiago de Chile Castillo Fontanaz/Direccion Meteorologica de Chile (2001: 379–382); Sao Paulo http://www.inmet.gov.br)

City	Climate type according to Köppen (1884)	Characteristics of temperature	Characteristics of precipitation	Characteristics of location
Bogotá	Temperate highland, tropical climate with dry winters	Mild temperature all year, little variation, annual average between 12.2 and 14.5 °C depending on location	Depending on location in the city: 600–1,200 mm/ 1,600 mm in Curubital	Located in a basin (2,600 m), proximity to high Andes, influenced by ENSO
Buenos Aires	Humid subtropical climate	Hot summers and mild winters (average maximum in the coldest month 7 °C)	Rainy all year, higher precipitation in the summer	Coastal zone, no elevation, influence of wind
Lima	Desert climate, extremely dry	warm temperature all year, little variation, annual average approx. 19 °C	Almost no precipitation (annual average <10 mm)	Coastal zone, strong influence of Humboldt and ENSO
Mexico City	Temperate highland, tropical climate with dry winters	Mild, annual average 15.6 °C, monthly average >12.9 °C	Rainy season in the summer (June–September >100 mm per month)	Located in a basin (2,500 m), proximity to high mountains
Santiago de Chile	Mediterranean, dry summer, subtropical	Annual average approx. 14 °C, differences between summer (average daily max >25 °C November–March) and winter (average daily min <5 °C May–August)	Extended dry season during summer, precipitation mainly in winter, total rainfall oscillates between 200 and 500 mm per year	Located in a basin, proximity to high Andes
Sao Paulo	Humid subtropical climate	Average daily maximum >20 °C per month	Rainfall all year, especially rainy season June-October (>100 mm per month)	

humid subtropical climate with rainfall throughout the year. The location of the cities is highly relevant, with some influenced by the Atlantic Ocean and others by the Pacific Ocean. Those on the Pacific Rim are affected by the Humboldt Current, and on a global scale by the El Niño Southern Oscillation (ENSO). The extremes of this climate pattern, El Niño and La Niña, lead to severe weather events.

Another striking element is the topographical location of the cities on a regional scale: coastal cities (Lima and Buenos Aires) contrast with those located in proximity to high mountains, as in the case of Santiago, Bogotá and Mexico City. This

Table 11.2 Expected climate changes and related impacts in the six Latin American megacities (*Source*: Krellenberg et al. (2013a: 35), based on RLN workshops. Data on changes in temperature and precipitation, and projection time frames: Bogotá Pabón Caicedo (2003: 114), and Krellenberg et al. (2013a: 34); Buenos Aires Camilloni (2009); Lima Krellenberg et al. (2013a: 34), and MINAM (2010); Mexico City INE (2012); Santiago de Chile Cortés et al. (2012); Sao Paulo Nobre (2010: 10))

City	Temperature	Precipitation	Extreme weather events	Impacts
Bogotá (by 2100)	Increase in average annual temperature of up to 4 °C (trends project increase up to 2.5 °C)	Reduction of up to 50 % (trends project reduction of 35 %)	La Niña: extreme rainfalls	Floods
			El Niño: droughts, water shortage	Landslides
				Fires
				Water scarcity
Buenos Aires (by 2080–2089)	Increase in average annual temperature between 0.6 and 1.9 °C (B2 scenario), between 0.9 and 2.8 °C (A2 scenario), min. temp. are expected to rise more than max. temp.	Slight reduction over next two decades, later increase of 12.5 % (B2 scenario)	Increase in storm events, higher intensity and extension over time	Floods
Lima	Increase in average annual temperature between 1 and 2–3 °C	Likely to stay extremely dry, but occasionally higher precipitation in upper Rio Rímac region	La Niña: extreme rainfalls possible	Floods
	Local climate may acquire more "tropical" characteristics			Landslides
				Urban heat Island effect
Mexico City (2080)	Increase in average annual temperature between 2 and 4 °C	Reduction in precipitation of up to 20 %	Very brief intense rainfall Increase in extremely dry periods	Floods
				Pollution
				Urban heat Island effect
				Droughts
				Water shortage
Santiago de Chile (2045–2065)	Increase in average annual temperatures, maximum temperature rising up to 2.5 °C	Highly uncertain	Reduction in the number of days with intense rainfall	Floods
		Decrease in rain days	Increase in precipitation with high temperatures	Water shortage
		Reduction in streamflow	Increase in number of days with max. temp. over 30 °C	
Sao Paulo (2030–2040/2050–2060/2080–2090)	Increase in average annual temperatures	Increase notably in heavy rainfalls	Increase in episodes of intense rainfall (>100 mm, rainfall events lasting at least 5 days)	Floods
		At the end of the century: more droughts		Landslides
				Urban heat Island effect

feature is highly significant when it comes to downscaling information from Global Circulation Models (GCM) to the metropolitan scale, since the mountains close to the metropolises concerned are not adequately represented in dynamic models.

The location of the cities in different climate zones also points to specific climate changes in the future. Based on a range of different methodologies and data sources, climate projections are available for all six metropolitan regions. The apparent differences should be taken into account and communicated with care.

Climate change modelling is generally related to a series of uncertainties: the underlying scenarios (IPCC 2000) make assumptions about socio-economic, political and technical development paths in order to estimate greenhouse gas emissions. These projections of global climate change include margins of probability, since GCM are themselves subject to uncertainties. The process of downscaling GCM output to the metropolitan scale—both the dynamic and statistical downscaling models—includes simplifications and occasionally data gaps, increasing the uncertainty of the results (see Cortés et al. 2012 and Chap. 2 in this book).

In brief, downscaling climate change to the urban-regional level is a contentious issue: although the data is effectively uncertain—even the direction of change (e.g., in precipitation) remains uncertain—action must be undertaken now to confront climate change, as ex-post adaptation is costly (Stern 2007) and, more importantly, can endanger lives and livelihoods. Regardless of uncertainty, therefore, it is vital to the fostering of informed decision-making to gain a comprehensive picture of the knowledge currently available on climate change at the metropolitan level.

Table 11.2 provides a summarized comparison of the principal climate changes to be expected in the six megacities under review, based on temperature and precipitation data. The very different methods of downscaling (cf. Krellenberg et al. 2013a, b) used to identify these changes renders them almost incomparable. They nonetheless constitute a valuable data base for each city in identifying expected climate change impacts and the appropriate adaptation strategies as response. Several common challenges referring to dynamic modelling were identified, such as the adequate representation of major mountain ranges (Bogotá, Mexico City and Santiago). The extreme weather events shown in the Table call for immediate response. Table 11.2 likewise gives an overview of the expected impacts of climate change, an impact assessment summary carried out by the Regional Learning Network.

As indicated in Table 11.2, the six selected cities share several aspects of expected future climate change and its related impacts, such as the obvious increase in temperature in all cases and an altered precipitation regime. This is consistent with their climatic zone. The key issue pertaining to extreme events is the amount of rain that falls within a short period, which produces the risk of high-impact floods and landslides, notably in mountainous cities. Furthermore, rising temperatures combined with falling precipitation rates, as in the case of Santiago de Chile, Mexico and Bogotá, will most likely result in more droughts and/or water scarcity (cf. Chaps. 2, 4 and 6 for Santiago de Chile).

11.3 The Need for Adaptation to Climate Change in Latin American Metropolises

After more than 20 years of intense research on and political attention to climate change by reducing greenhouse gas emissions, the focus has begun to shift towards adaption to this phenomenon, especially since the 2000s (Bulkeley 2010). Since climate change is a given (IPCC 2007), societies must adapt to the changes already visible today. Mitigation measures, albeit indispensable, are a mere drop in the ocean, however, when it comes to confronting the adverse effects of climate change. For this reason, adaptation should be seen as a central issue of climate change policy (Satterthwaite et al. 2007). The impact of flooding and landslides in particular calls for policies that refer to risk prevention, vulnerability reduction and the fostering of resilience.

In the context of highly segregated cities such as those in Latin America under analysis, any element of risk or exposure is potentially a social challenge and should be considered in the context of land-use planning, housing policies, urban transportation and health services. The discussion on climate change adaptation in the large Latin American cities is thus closely related to the issue of sustainable development and vulnerability. The IPCC (2007) has also referred to this linkage, one that has been widely discussed in the context of co-benefits. Scholars working on climate change in the Latin American context frequently request in-depth conceptual discussions on the issue of vulnerability in order to tackle the connection between exposure and impact conceptually (Lampis 2012; Romero Lankao and Qin 2011). This issue has likewise given rise to criticism of the excessive use of a fashionable but at times fuzzy concept of vulnerability (Sánchez-Rodríguez 2011). Nevertheless, current adaptation discourses concentrate frequently, albeit not exclusively, on risk reduction in the case of floods and storms (Chafe 2007). Urban flooding is a common occurence in low-elevation coastal zones or where the specifics of urban locations with increased construction activity show lack of natural retention areas and the technical infrastructure to cope with extreme rainfall events (Huq et al. 2007; Krellenberg et al. 2013a; Young 2013).

Short-term solutions are a priority here if risks are to be reduced and future hazards prevented. The very nature of climate change, however, is clearly long-term and requires the establishment of long-term horizons in metropolitan planning and strategy development. Although heat waves and scarcity of water resources, for example, affect the entire city, their spatial shape and their impact on households and individuals are closely related to spatial patterns of vulnerability and social inequality (cf. Chap. 7). This means long-term policies that link climate change adaptation to urban land-use planning and risk management.

The debate on urban adaptation to climate change must take a wider spatial context into account than cities, since a number of adverse effects can occur beyond the urban centres, such as food security or water resources that rely on watersheds. The impact of climate change outside the metropolitan area can have an indirect effect on cities in social and economic terms. This perspective on climate change refers primarily to problem-solving and policy and territorial planning, all of which emerge, for example, when a demand for coordinated action arises in the context of water resources. In the case of Colombia (Bogotá) and the surrounding region of Cundinamarca, planning activities show evidence of this concern. The same holds true for Santiago de Chile (see Chap. 4).

11.4 Climate Change Policy in Latin American Cities in a Rapidly Changing Context

This section discusses the prevailing public policies on climate change in the six selected cities. It includes elements that stem from national and urban-regional scales, such as climate change plans, strategies and programmes emanating from national negotiations or bottom-up initiatives associated with administrative policy. Recent years have seen the emergence of a wealth of documents dealing with climate change, some of which were formalized, and the approval of a number of laws. In many cases, response translates to work in progress, since public policy is currently undergoing a dynamic phase of rapid change, all of which makes the assessment of existing policies a formidable challenge at this stage, if not impossible. Consequently, information presented in this section does not claim to be comprehensive but rather summarizes what is happening in general terms in the six selected cities and looks at agenda topics, participating institutions and the process itself.

To gain a more coherent understanding of ongoing response, plans, strategies and programmes at urban-regional level and of those provided as a framework at the respective national level are presented in the following, city by city. Key elements of individual climate change actions are introduced and features considered of particular relevance to other cities highlighted, e.g., successful negotiations or concrete measures in one city that are applicable to other cities confronting similar administrative challenges.

11.4.1 Mexico City and Mexico

A National Climate Change Strategy was defined in Mexico in 2007 ("Estrategia Nacional de Cambio Climático"). Since Mexico has a federal constitution, its state level has the power to decide upon and implement a climate change agenda.

Of the Latin American megacities under review, Mexico City seems to have the most advanced and most convincing climate change policy. Following the presentation in 2006 of a Strategic Plan for Local Action to climate change, a more concrete Climate Action Plan was introduced in 2008 ("Plan de Acción Climatica Ciudad de México 2008–2012"). The Plan is comprehensive and includes a broad range of actions with clearly defined goals, responsibilities and schedules. Although the Action Plan itself focuses almost entirely on mitigation measures, it can be interpreted as "best practice" in terms of climate action in Latin American megacities. Furthermore, a law on Mitigation and Adaptation to Climate Change and Sustainable Development was approved in 2011. These—undeniably positive—examples of real climate action sustained by a combination of political will, priority of environmental issues on the political agenda, and academic and technical knowledge suffer from one major constraint: plans and laws refer to the jurisdiction of the Federal District (México D.F.), thereby limiting the Climate Action Plan to a somewhat arbitrarily divided section of the metropolitan area. Buenos Aires, which will be discussed later on, shows evidence of comparable circumstances.

11.4.2 Sao Paulo and Brazil

Similarly challenging is the situation in Brazil, due to its administrative structure. Three administrative levels must be taken into account here, all of which are mandated to take the relevant climate change action, including the adoption of decrees and laws: the national level, the Sao Paulo state level and the local municipal level. The National Plan on Climate Change (based on a decree from 2007) and the National Policy on Climate Change ("Política Nacional sobre Mudança do Clima, PNMC") exist at national level. The state of Sao Paulo has defined a State Programme on Climate Change (PROCLIMA) and a Policy on Climate Change of the State of Sao Paulo ("Política Estadual de Mudanças Climáticas, PEMC"). Outstanding here is the municipal law on climate change (Ley Nº 14.933, June 2009) adopted at local level and the "Guidelines for the action plan of the city of Sao Paulo to mitigate and adapt to climate change" based on an initiative by the mayor of Sao Paulo.

11.4.3 Buenos Aires and Argentina

In Argentina, a National Strategy on Climate Change is in the process of elaboration (since 2010) although unmistakable progress has been made at the provincial level of Buenos Aires. In the case of the federal capital, Buenos Aires, an Action Plan has been formulated for the year 2030 ("Plan de Acción Buenos Aires 2030"), with an obvious bias towards mitigation measures. A key concern is the administrative separation of the federal capital of Buenos Aires (Ciudad Autónoma de Buenos Aires) from the province of Buenos Aires, dividing the Metropolitan Region (Gran Buenos Aires) into the inner circle, on the one hand, and the outer circle and suburban neighbourhoods, on the other, where more than 50 % of the population of Buenos Aires resides. Hence the potential impact of this Action Plan referring to the federal capital only is heavily confined.

In September 2011, a law on Adaptation and Mitigation to Climate Change was approved, Ley 3.871 (http://www.cedom.gov.ar/es/legislacion/normas/leyes/ley3871.html). It prescribes implementation of mitigation and adaptation measures, including the issue of vulnerability, and explicitly calls for cooperation between the federal capital and the province.

11.4.4 Bogota and Colombia

Colombia is currently (2011–2013) working on a National Plan for Adaptation to Climate Change. Several institutions at metropolitan level are engaged (in 2012) in designing an integrated regional plan for climate change adaptation for the Bogotá-Cundinamarca region ("Plan Regional Integral de Cambio Climático" PRICC Región-Capital). The Capital Region initiative, which comprises the city of Bogotá and the surrounding region of Cundinamarca, is remarkable in so far as it embraces integration in concrete terms, in contrast to Buenos Aires and Mexico City, still struggling with legal constraints on their action plans. It should be stressed, however, that—unlike these two cities—a high percentage of the total population of the Metropolitan Region resides in the capital district Bogotá D.C. and that its administrative fragmentation is less challenging. Furthermore, Bogotá has a major political authority in the "alcalde mayor". Nevertheless challenges for a climate change agenda have been pointed out e.g. by Lampis (2008).

11.4.5 Santiago and Chile

Climate change policy in Chile concentrated primarily on the national scale until 2012. A National Strategy on Climate Change was elaborated in 2006 and led to the National Action Plan 2008–2012 ("Plan de Acción Nacional de Cambio Climático 2008–2012"). Notwithstanding this unambiguous political statement outlining fields of action, its goals are unclear, it displays a marked sector divide, makes little reference to adaptation and lacks an urban focus. Conspicuous in the case of Chile is that the most successful action has been taken in sectors where the process of adapting to adverse effects of climate events is an inherent mechanism, such as in the agricultural sector. Agriculture has a rich tradition of preventing and handling climate-related damage to production and is a notable example of how climate change adaptation can be realized on the national scale by linking the measures to sectorial policies. Conversely, this indicates that the absence of robust sectorial support could frustrate the implementation of adaptation strategies. At urban level, the Regional Climate Change Adaptation Plan adaptation (cf. Chaps. 9 and 12) was developed with the participation of a large group of local stakeholders and handed over at the end of 2012 to the regional authorities responsible for climate change.

11.4.6 Lima and Peru

A National Strategy on Climate Change was first formulated in Peru in 2003 and renewed in 2009. This was followed by the adoption of an Action Plan for Climate Change Adaptation and Mitigation ("Plan de Acción de Adaptación y Mitigación frente al Cambio Climático") in 2010. These instruments have so far failed to lead to concrete action of any consequence. The metropolitan area of Lima has made little progress. Only in 2012 did Lima begin to work on a Metropolitan Strategy for Climate Change.

11.5 Adaptation to Climate Change: A Classification of Action in Latin American Megacities

Most of the climate change plans and programmes of the six cities clearly lack defined action in the field of adaptation. This is due in essence to the generic character of these documents and their emphasis on mitigation measures. Nevertheless, a number of adaptation measures have been undertaken, some of which are presented in this section. Priority is given to action directly related to adaptation, although it may not be part of a formal adaptation plan. In many instances, adaptation to climate change calls for action that is already an essential ingredient of sectorial policies aimed at sustainable urban development. Sectorial policies are mentioned here in the context of adaptation measures for methodological purposes, since co-benefits from climate action and sustainability policies are a compelling opportunity, not merely for synergetic action but also in terms of a political agenda that will take "big issues" into account and lead to similar or complementary measures and projects.

The concrete adaptation measures identified in the six cities under analysis are frequently related to *risk prevention and disaster management*. As shown earlier, extreme weather events, notably heavy rainfall, may have an immediate impact on urban society in the form of floods and landslides. Hence risk prevention and disaster management—although not explicitly related to climate change—can be seen as highly relevant adaptation measures. In the six cities, two such measures can be distinguished: (1) In the case of Sao Paulo, land-use planning and relocation programmes to remove informal settlements from risk areas take high priority in climate change adaptation, although originally linked to housing policy and urban development. (2) In Bogotá and Colombia, risk reduction and disaster management have attracted substantial political attention and led to appropriate action in recent years. The Risk Management Handbook for Municipalities, a promising institutional structure, was introduced to connect the national (*Unidad Nacional para la Gestion del Riesgo de Desastres* UNGRD) with the local level. Importantly, it relates to the National Plan for Adaptation to Climate Change, which in turn is linked technically to the Meteorological Service (IDEAM) to facilitate early warning of extreme weather events and landslide-prone situations.

A second area of action is *water management* even if not necessarily explicitly related to climate change (see for instance the case of Sao Paulo in Silva 2008). Hydrological cycles in most of the cities under review have been altered by climate change; seasonal patterns and water supplies are most likely to undergo further change in the future. At the same time, urban population growth and increased prosperity has led to a greater demand for water and hence to a shift in domestic water-use patterns, which in turn affects river watersheds and water body recharge. Adapting to this pressure could enhance the supply of drinking water with, for example, technical measures to reduce water loss (Mexico City and Santiago), long-term technical solutions such as sea water salinization (a reality in northern Chilean cities) or water transport from distant water reserves (Mexico City and Sao Paulo). Some action has been taken on the water demand side, as in Lima, where reduction of per capita water consumption is financially rewarded.

Another relevant issue is *(public) health*, with reference to the increase in the number and intensity of heat waves. The spatial occurrence of heat islands depends on land-use types and thus coincides to some extent with socially segregated land-use patterns, as in the case of low-income urban areas with few green spaces (cf. Chap. 6). It can be assumed that future heat waves will affect the poorer population to a greater degree. Residents of this social group are less favoured with health service coverage and have fewer options for adaptive measures in their immediate environment, such as infrastructure investments, ultimately making the health impacts of climate change in Latin American cities a social problem. The adaptation to climate change so urgently needed is also related to other policy fields: (a) housing policy and construction norms, (b) land-use planning—notably public space—and, finally, (c) public health services. The quality of buildings is seen as having considerable potential in terms of mitigation. It likewise refers to adaptation, since heat exposure can be reduced by the use of appropriate construction materials, for example, and green walls in the case of offices or public buildings. This in turn enhances mitigation policies as a result of the co-benefits involved, for the most part a reduction in energy consumption. The implementation of green areas, eco-efficient buildings and construction norms for temperature isolation and energy efficiency are highly relevant actions. Several measures have a direct link to long-term urban planning, e.g., fostering a combination of high density and increased green spaces in housing areas.

Table 11.3 presents an overview of adaptation measures in the six cities under review, using classifications proposed by the European Environment Agency that distinguish between green infrastructure, grey infrastructure and soft, institutional measures (EEA 2012).

11.6 Conclusion

As the findings on the six selected Latin American megacities show, climate change strategies require a multidisciplinary approach and should include adaptation in a comprehensive and integrated manner. A striking example of the need for such a trans-sectorial approach is the field of land-use planning; air pollution and heat islands are key factors with an immediate link to land-use patterns and material flows (e.g., mobility). At the same time, land-use planning and housing standards can greatly reduce the impact of heat islands and heat waves on people's well-being and health.

In all of the metropolises under analysis, initial steps have been undertaken to institutionalize climate change action on the urban scale, albeit the approaches and experiences vary. Addressing adaptation within an existent institutional framework

Table 11.3 Summary of existing adaptation measures in the six Latin American megacities (*Source*: Krellenberg et al. (2013a: 79), based on RLN workshops, using classification of EEA (2012))

Metropolitan area	Grey measures (infrastructure)	Green measures	Soft measures (institutional)
Bogotá	Buildings with ventilation systems and passive cooling	Revitalization of city centre, including more green spaces	Socio-economic differentiation of water prices
			Urban planning to avoid occupation of slopes and low-lying areas on river banks
			Risk management, fire prevention campaigns
			Educational campaigns
Buenos Aires	Improved materiality of housing, reduction of its physical vulnerability	Green roofs, incorporation of trees into local politics	Warning systems and urban signage
	Implementation of storm sewer system		Promotion of energy efficiency
			Combat climate-related health risks and emergence of tropical diseases
Lima	Eco-efficient housing, passive cooling systems	Vegetation recovery (on slopes)	Right to water, major sections of society still without access
	Water desalination	*Lima verde* programme, gardens with xerophilous plants	Financial incentives to foster efficient water use
	Reduction of water loss, leakage and misuse		Cooperation with upper water catchment area
	Treatment plants, sewage and recycled water use for landscape irrigation	*Mi huerta* (my orchard) programme	Water law
			Construction standards
			Flexible working hours
			Changes to water cost structure (infrastructure versus consumption) and value assignment
			Combine territorial policies with land-use and water policy
Mexico City	Pedestrian zones		Adaptation strategy development
	Bike lanes		Discouragement of private car use
	Higher water efficiency with recycled grey water		Inclusion of climate change adaptation actions in urban development plans
			Drinking water access system for all socio-economic groups
			Green Mortgage (ecological housing incentive)

(continued)

Table 11.3 (continued)

Metropolitan area	Grey measures (infrastructure)	Green measures	Soft measures (institutional)
Santiago de Chile	Promotion of modern irrigation	Fewer pavements to reduce heat islands	Watershed management
	Efficient use of drinking water	Recovery of urban vegetation	Urban landscape development and green area plan
	Groundwater recharge, sea water desalination, wastewater recycling		Thermal regulation of construction
	Construction of sustainable public buildings		
	Public transport (exclusive lanes)		
Sao Paulo	Drain pools	Increase in permeable urban area	Holistic view of water management
	Efficient urban drainage system, rain water channelling		Multiple water use
			Plans and programmes to combat flooding; flood risk management

is possible, as seen in the case of Bogotá D.C. Taking these activities as a starting point, concrete action for climate change adaptation was developed. In Mexico City and Sao Paulo, for example, climate change plans were defined and put into practice, and later underpinned by specific climate change laws, thereby creating a sound institutional and legal basis for climate change action.

This notwithstanding, most of the measures in question are designed for climate change mitigation purposes rather than adaptation and are found to be more effective when linked to other strategic goals or national urban policies and thus to co-benefits. In the area of mitigation, for example, greenhouse gas reduction targets reinforce energy efficiency measures, enhance urban transport and promote the reduction of waste and pollution. Similarly, adaptation policies can generate major co-benefits, e.g., more efficient water use, construction and urban design for healthier communities and the implementation of risk management systems. These elements of climate change adaptation coincide with those employed to secure more sustainable cities without specific reference to climate change, leaving the impression that the exact dimension of climate change in the urban context has not yet acquired priority status.

Large Latin American cities are characterized by extreme income disparities, socio-spatial segregation and hitherto unsolved gaps in terms of access to certain goods and services, including basic needs (e.g., Kabisch et al. 2012). Although essentially an issue of sustainable development and equity, this context is also relevant for climate change adaptation, as the need for adaptation varies considerably within the city. Furthermore, most of the measures discussed are costly when it comes to improving construction, increasing green areas, implementing more efficient infrastructure or changing patterns of spatial behaviour of households and individuals may be obliged in this case to share the burden of direct or implicit costs. Given the uneven distribution of wealth in the six cities, major sections of the urban population will not be in a position to carry these costs, although they may well be among those who suffer most from the adverse effects of climate change. Hence income disparities should be addressed in adaptation plans, since the capacity of the population to embrace climate change action is heavily dependent on income.

Monitoring processes of adaptation in the metropolitan cities of Latin America to evaluate progress and learn from the success of others represents a useful tool for the implementation of climate change policies. The established Regional Learning Network (see Krellenberg 2013a) could be a step in this direction. It has been observed that representatives of the cities under review have begun to intensify their contacts and it can be assumed that they will exchange ideas and ultimately design measures in each city that are more specific and more effective.

References

Bulkeley, H. (2010). Climate policy and governance: An editorial essay. *Wiley Interdisciplinary Reviews: Climate Change, 1*, 1311–1313.

Camilloni, I. (2009). *Cambio Climático*. http://www.atlasdebuenosaires.gov.ar

Castillo Fontanaz, C. (2001). *Estadística Climatológica Tomo I*. Santiago: Dirección Meteorológica de Chile.

Chafe, Z. (2007). Reducing natural disaster risk in cities. In The Worldwatch Institute (Ed.), *State of the World 2007: Our urban future* (pp. 112–129). New York/London: WW Norten & Co.

Cortés, G., Schaller, S., Rojas, M., Garcia, L., Descalzi, A., Vargas, L. et al. (2012). *Assessment of the current climate and expected climate changes in the Metropolitan Region of Santiago de Chile*. Research Report Climate Adaptation Santiago (CAS).

EEA (European Environment Agency). (2012) *Urban adaptation to climate change in Europe*. Challenges and opportunities for cities together with supportive national and European policies. EEA Report 2/2012. http://www.eea.europa.eu/publications/urban-adaptation-to-climate-change

Heinrichs, D., Aggrawal, R., Barton, J., et al. (2011). Adapting cities to climate change: Opportunities and constraints. Findings from eight cities. In D. Hoornweg, M. Freire, M. Lee, P. Bhada-Tata, & B. Yuen (Eds.). *Cities and climate change: An urgent agenda*. Washington, D.C.: The International Bank for Reconstruction and Development.

Huq, S., Kovats, S., Reid, H., & Satterthwaite, D. (2007). Editorial: Reducing risks to cities from disasters and climate change. *Environment and Urbanization, 19*(1), 3–15.

IIED (International Institute for Environment and Development). (2007). Reducing risks to cities from climate change: An environmental or a development agenda. *Environment and Urbanization Brief, 15*, 1–6.

INE (Instituto Nacional de Estadísticas). (2012). *El cambio climático en México*. Información por estado y sector. Available at: http://www.ine.gob.mx/cclimatico/edo_sector/estados/futuro_mexico.html

INEI (Instituto Nacional de Estadística e Informática). (2007). *Perú Compendio Estadístico Nacional 2007*. Lima.

IPCC. (2000). *Informe Especial del IPCC*. Escenarios de emisiones. Resumen para responsables de políticas. Available at: http://www.ipcc.ch/pdf/special-reports/spm/sres-sp.pdf

IPCC. (2007). Cambio climático 2007: Informe de síntesis. In Equipo de redacción principal: Pachauri, R. K. y Reisinger, A. (directores de la publicación) *Contribución de los Grupos de trabajo I, II y III al Cuarto Informe de evaluación del Grupo Intergubernamental de Expertos sobre el Cambio Climático*. Geneve: IPCC.

Jordán, R. (2008). The responses of urban areas to climate change: Perspectives on energy efficiency in megacities of Latin America. *UGEC Viewpoints, 1*, 17–21.

Jordán, R., Rehner. J., & Samaniego, J. (2010). *Regional Panorama Latin America. Megacities and sustainability*. Santiago de Chile: CEPAL Project Documentation.

Kabisch, S., Heinrichs, D., Krellenberg, K., Welz, J., Rodriguez, J., Sabatini, J., et al. (2012). Socio-spatial differentiation: Drivers, risks and opportunities. In D. Heinrichs, K. Krellenberg, B. Hansjürgens, & F. Martínez (Eds.), *Risk Habitat Megacity* (pp. 155–181). Berlin: Springer.

Köppen, W. (1884) Die Wärmezonen der Erde nach der Dauer der heißen, gemäßigten und kalten Zeit und nach der Wirkung der Wärme auf die organische Welt betrachtet. In: Meteorologische Zeitschrift (Vol. 1, pp. 215–226) (The thermal zones of the earth according to the duration of hot, moderate and cold periods and to the impact of heat on the organic world. transl., ed. E. Volken, S. Brönnimann. (2011). *Meteorologische Zeitschrift 20*, S.351–S.360. doi: 10.1127/0941-2948/2011/105).

Krellenberg, K. (2012). A science-policy approach towards local adaptation planning: The case of Santiago de Chile. In K. Otto-Zimmermann (Ed.), *Resilient cities 2 - Cities and adaptation to climate change – Proceedings of the global forum 2011* (pp. 233–240). Dordrecht: Springer.

Krellenberg, K., Jordán, R., Rehner, J., et al. (2013). *Adaptation to climate change in megacities of Latin America. Regional Learning Network of the research project ClimateAdaptation-Santiago (CAS)*. Santiago de Chile: CEPAL Serie Seminarios y Conferencias.

Krellenberg, K., Müller, A., Schwarz, A., Höfer, R., & Welz, J. (2013b). Flood and heat hazards in the Metropolitan Region of Santiago de Chile and the socio-economics of exposure. *Applied Geography, 38*(2013), 86–95.

Krugman, P. (1997). *Development, geography, and economic theory*. Cambridge: MIT Press.

Lampis, A. (2008). *Urbanization and global climate change: The case of Bogotá*. UGEC Viewpoints No 1, Integrative Perspectives on Urbanization and Climate Change. A focus on the Americas. Available at: http://www.ugec.org/docs/UGECAugust08.pdf

Lampis, A. (2012). Risk, vulnerability and adaptation capacity: A reflection about the non-reducibility of social and bio-physical vulnerability. *GLP News, Newsletter of the Global Land Project 8*, 24–26. Available at: http://www.globallandproject.org/arquivos/GLP_news_march_2012.pdf

MINAM (Ministerio del Medio Ambiente). (2010). *Segunda comunicación nacional del Perú a la CMNUCC*. Lima: MINAM.

Nobre, C. A. (2010). *Vulnerabilidade das Megacidades Brasileiras às Mudanças Climáticas: Região Metropolitana de São Paulo. Sumário Executivo.*

ONU Habitat. (2012). *Informe mundial sobre asentamientos humanos 2011. Las ciudades y el Cambio Climático: Orientaciones para políticas. Resumen Ejecutivo*. London/Washington, DC: Earthscan.

Pabón Caicedo, J. D. (2003). El cambio climático global y su manifestación en Colombia. *Cuadernos de Geografía, 12*, 111–119.

Romero Lankao, P. (2007). Are we missing the point? Particularities of urbanization, sustainability and carbon emissions in Latin American Cities. *Environment and Urbanization, 19*(1), 159–175.

Romero Lankao, P. (2010). Impacts, adaptation and vulnerability to global environmental change: Challenges and pathways for an action-oriented research agenda for middle-income and low-income countries. *Current Opinion in Environmental Sustainability, 2*, 364–374.

Romero Lankao, P., & Qin, H. (2011). Conceptualizing urban vulnerability to global climate and environmental change. *Current Opinion in Environmental Sustainability, N° 3*, 142–149.

Sánchez-Rodríguez, R. (2011). Urban and social vulnerability to climate variability in Tijuana, Mexico. In R. E. Kasperson & M. Berberian (Eds.), *Integrating science and policy* (Vulnerability and resilience in global environmental change, pp. 187–211). London/Washington, D.C.: Earthscan.

Satterthwaite, D. (2006). *Climate change and cities. Sustainable development opinion*. London: IIED.

Satterthwaite, D., Huq, S., Pelling, M., Reid, H., & Romero Lankao, P. (2007). *Adapting to Climate Change in Urban Areas. The possibilities and constraints in low- and middle-income nations. Theme: Climate Change and Cities 1*. IIED Human Settlements Discussion Paper Series.

Silva, T. R. (2008). Total urban water management in São Paulo, Brazil: A case of ongoing experiences on applied research and public policy. *UGEC Viewpoints, 1*, 26–29.

Stern, N. (2007). *The economics of climate change. The stern review*. Cambridge: Cambridge University Press.

The World Bank. (2009). *World development report 2009. Reshaping economic geography*. Washington, D.C.: The World Bank.

UN ECLAC. (2012). *Population, territory and sustainable development*. Santiago de Chile.

UN DESA. (2008). *World urbanization prospects: The 2007 revision*. New York: United Nations.

Young, A. F. (2013). Urbanization, environmental justice, and social-environmental vulnerability in Brazil. In C. G. Boone & M. Fragkias (Eds.), *Urbanization and sustainability. Linking urban ecology* (Environmental justice and global environmental change. Human-environment interactions, Vol. 3, pp. 95–116). Heidelberg/New York/London: Springer.

Developing a Regional Climate Change Adaptation Plan: Learning from Santiago de Chile

12

Kerstin Krellenberg, Katrin Barth, and Bernd Hansjürgens

Abstract

This chapter draws conclusions and discusses lessons learned from Santiago de Chile by evaluating the overall process that led to the Regional Climate Change Adaptation Plan. Highlighting both the benefits and constraints of the applied approach, it takes a closer look at the various steps undertaken throughout the entire chain of analysis. It re-examines the actors involved and reflects on the role of science in climate change adaptation. The chapter furthermore addresses appropriate implementation and transfer mechanisms.

Keywords

Regional Climate Change Adaptation Plan • Implementation • Process evaluation • Transdisciplinarity

12.1 Climate Change Adaptation as a Complex Process

As argued in Chap. 1, obstacles to and constraints on climate change adaptation plans, strategies and measures under real-world conditions are a frequent occurrence and in many cases hamper successful implementation. Accordingly, the practical experience of developing and implementing climate change response is of the utmost importance and of particular benefit to the relevant actors (decision-makers, scientists and civil society), whose role it is to design and implement climate change adaptation strategies.

K. Krellenberg (✉) • K. Barth
Department of Urban and Environmental Sociology, Helmholtz Centre for Environmental Research – UFZ, Permoser Str. 15, 04318 Leipzig, Germany
e-mail: kerstin.krellenberg@ufz.de

B. Hansjürgens
Department of Economics, Helmholtz Centre for Environmental Research – UFZ, Permoser Str. 15, 04318 Leipzig, Germany

K. Krellenberg and B. Hansjürgens (eds.), *Climate Adaptation Santiago*,
DOI 10.1007/978-3-642-39103-3_12, © Springer-Verlag Berlin Heidelberg 2014

This final chapter draws some general conclusions from practical lessons learned in Santiago de Chile, focusing on the triumphs (and difficulties) in the development, prioritization, selection and implementation of climate change adaptation measures and their subsequent embedment in a Regional Climate Change Adaptation Plan (RCCAP). The foregoing chapters exemplified the different steps of the underlying concept of an integrative inter- and transdisciplinary approach (IIT) as presented in Chap. 1, the key features of which were the participatory process and a mutual learning network of Latin American mega-agglomerations. Scientific analysis, the mainstay of the participatory process and the development of adaptation measures, clarified the complexity of urbanization, its interrelation with future climate change, the expected consequences, and climate change adaptation for the Metropolitan Region of Santiago de Chile (MRS), concentrating on such pressing issues as water and energy supplies, as well as flood and heat hazard and their related vulnerabilities.

In order to evaluate the overall success of the adopted approach, this final chapter examines the steps undertaken in the "chain" of analyses: regional climate change (Chap. 2), climate change impact analysis (Chaps. 4, 5, 6, and 7), and adaptation needs and measures (Chap. 9) at urban-regional level. Accordingly, it re-examines the process of strategy development and implementation, concentrating on actor experiences and opinions as a means of assessing the inter- and transdisciplinary process. Starting point for the analysis was participant feedback from roundtable meetings, which was effected via an online questionnaire (cf. Chap. 10), an evaluation tool that allows for additional conclusions on the benefits and constraints of the entire process.

Based on this information, the remaining sections of the closing chapter address the following aspects:

- Section 12.2 focuses on lessons learned in the participatory process with a discussion on the relevant benefits and constraints.
- Since the integrative inter- and transdisciplinary research approach presented in this book is science-initiated and science-driven, Sect. 12.3 reflects on the role of science in such processes.
- Taking into account that the scientific part of the process culminated in handing over the Regional Climate Change Adaptation Plan to the local authorities in Santiago de Chile and implementation is ongoing, Sect. 12.4 briefly discusses appropriate implementation mechanisms.
- Coming back to the Latin American perspective presented in Chap. 11, Sect. 12.5 concludes with an outlook on possible transfer mechanisms.

12.2 Reflections on the Inter- and Transdisciplinary Process

Apart from the methodological constraints that emerged in the course of downscaling global climate models (see Chap. 2) and the climate change impact analysis (see Chaps. 4, 5, 6, 7 and 8), including explorative scenario analysis according to Chap. 3 the main challenge to developing the Regional Climate Change Adaptation Plan was the participatory process, the transdisciplinary component of the underlying procedure (see Chap. 9). The science-policy dialogue was faced with the task of bringing scientists and stakeholders from numerous organizations and disciplines together and jointly developing an adaptation plan in spite of poor awareness of climate change, knowledge gaps and divergent stakeholder interests. Having analysed practical experiences and the responses to the online questionnaire, the following considerations seem worthy of attention:

1. Climate change adaptation is frequently confronted with lack of awareness about climate change and its associated impacts, and the need for response. A remarkable lack of awareness of the subject also prevailed in the MRS—among decision-makers, civil society and indeed scientists, since no detailed analysis existed—making the task of finding the best way to establish the topic and make it a high priority on the political agenda a tough challenge from the outset of the process. Hence, vast efforts were devoted to closing this awareness gap by first of all elaborating a sound scientific data base on regional climate change and its impacts for the MRS, and secondly, providing roundtable participants with information in an appropriate manner. Of vital importance was the timely communication of the scientific work on climate change and its expected impacts for the MRS to local stakeholders. This was especially challenging as roundtable meetings took place from the very beginning of the process, parallel to (and not after) the scientific analysis, in order to guarantee early, comprehensive stakeholder involvement. Although final scientific results on climate change and its impacts in Santiago de Chile were not ready for communication at the beginning of the process, general information on adaptation, urban planning and scenarios, for example, was provided at roundtable meetings. This was key to holding stakeholder interest. As documented in the stakeholder responses to the online questionnaire, particularly information on the experience of other cities in the world attracted great interest and helped to increase knowledge and awareness.
2. Good communication is key for transdisciplinary approaches. General information on climate change and climate adaptation or the specific results for the situation in the MRS presented and discussed in the roundtable meetings was made available in the form of briefing and working papers. In this context, engineering a method of communicating scientific information and results in the interests of reaching a "common language and understanding" between scientists and local actors was a formidable task. Vast efforts were made to provide the results in Spanish in a manner likewise comprehensible for non-experts in the various fields. Looking at the stakeholder feedback, this seems to have been successful. All of the online questionnaire respondents indicated that the exchange of knowledge between science and practice had worked to their satisfaction. They furthermore stated that intercontinental communication—the process coordination group and half of the scientists were German, while the other half were based in Chile—also functioned satisfactorily, with language and cultural differences presenting no further problem. In addition, it should be stressed that all stakeholder respondents found the email provision (or print versions) of project material highly useful, whereas scientific publications available in English on the website were accessed to a substantially

lower degree. This clearly underpins that the scientists were obliged to consider channels, products and instruments other than scientific articles in order to communicate their results to the different target groups involved.

3. Roundtable meetings need to be structured flexibly. Apart from the presentation of scientific findings, participating stakeholders were invited to air their own views on the context-based topics. This led to a knowledge match of science and local practice, and opened up new avenues for discussion. Stakeholders deliberated over science-related information and contributed actively to the elaboration and implementation of the selected adaptation measures.—It was precisely this mix of inputs (presentations and discussions) that reaped positive feedback from the polled stakeholders. In their view, coupled with a variety of moderation methods these elements worked to advantage.
4. Interestingly, although stakeholders confirmed the importance of the science-practice dialogue for access to scientific climate change data, it was not the cardinal motivation for their participation in the process. What drove them most was their interest in establishing and enlarging a common knowledge base. Their stakeholder needs were satisfied by "what" was communicated and "how". There was a consensus that the results presented and discussed had great practical relevance and provided new information highly pertinent to Santiago de Chile. It can be seen as an overall achievement that participants indicated they had shared roundtable discussions with their organizations (e.g., roundtable minutes, conversations with colleagues, presentations, university teaching) and thus contributed to disseminating the knowledge they acquired.

Stakeholders also made a number of crucial remarks that deserve attention, including acknowledgement of the up-to-date scientific data base used in the analysis. Although not explicitly mentioned, it can be assumed that this statement refers primarily to flood hazard maps (see also Chap. 6). The stakeholders spoke of more recent data but did not make it available. Another critical point in their view was the somewhat marginal treatment of the rural areas of the MRS in the analysis, and that the integration of aspects of social behaviour would have been to its advantage. Since time and resource constraints did not allow for the inclusion of such aspects during the process, these useful comments should be carefully considered in any future activities.

Most questionnaire respondents (87 %) agreed with the choice of participating institutions. Here, a number of them would like to have seen a more significant presence of national actors, personnel with greater powers of decision, and public services. Changes in job positions proved to be challenging throughout, as several key participants at roundtable meetings were either replaced or simply dropped out. This high fluctuation of individuals (not institutions) put a sizeable strain on the overall organization of the process. The fast rhythm of political agendas and the attendant topics was similarly challenging. Each roundtable meeting required an update to ensure a common understanding among the participants. Changes in personnel called for clarification and discussion of certain topics several times. It is worth mentioning in this context that the duration of each roundtable meeting (three hours) and the time distance between these events (approx. three to five months) was seen by stakeholders as somewhat problematic. A request for higher frequency was made a) to have sufficient time to discuss the proposed measures in greater detail and b) to reduce the period of time between each roundtable event.

In sum, the outcome of the feedback questionnaire reveals that the overall process achieved a change in attitudes and made a significant contribution to knowledge on climate change and adaptation. The success of the approach was endorsed by the official ceremony that marked the submission of the Regional Climate Change Adaptation Plan to the key political entities responsible for its implementation.

12.3 Evaluating the Role of Science

The problem statement (lack of climate change response at local level), the initiative (application for funding in Germany), and the development of an overall project structure (design of the integrative inter- and transdisciplinary approach) as presented in this book were science-driven and subsequently organized and executed by a German scientific coordination team. This notwithstanding, the project first got under way after a Memorandum of Understanding had been signed, legally and institutionally endorsing the strong commitment of the key Chilean groups responsible for implementation of the Regional Climate Change Adaptation Plan: Gobierno Regional (GORE) and Regional Secretary of the Environmental Ministry (SEREMI MA).

With regard to the role of science in Chile it can be said that scientific work is traditionally used to validate political decisions by giving legitimacy, for example, to regional data and analysis in highly sector-divided areas (see Friedmann and Stöhr 1966). Hence it was more conventional in the past for science to identify problems rather than to provide solutions in terms of political action. Historically anchored in Chile is likewise the importance of personal contacts and networks dating back to high school or university which facilitate the communication when it comes to decision-making (Raul Urzua 2000).

Although not renowned for close collaboration, the Chilean scientific community is interconnected and more or less centralized in the city of Santiago (Ramon Latorre 2001). Given the absence of a coordinated policy for the development of national science capacities, there is little official cooperation between the different institutes and universities (Hansen et al. 2002; Holm-Nielsen and Agapitova 2002).

Scientific experts and decision-makers know each other nevertheless und such contacts are availed of informally for policy advice (Raul Urzua 2000). Bearing this in mind, the participatory process of developing the Regional Climate Change Adaptation Plan did break new ground.

Furthermore, the entire process took place against the backdrop of existing institutional and legal frameworks. In other words instead of waiting for overall political change to push for climate change adaptation and pave the way for a Regional Climate Change Adaptation Plan for Santiago, the aim of the approach was to deal flexibly with the given. The idea was to find a niche that would raise the topic of “climate change adaptation” on the political agenda (cf. Krellenberg and Barth 2014). It could therefore be argued that the strong commitment of GORE and SEREMI MA to the science-driven process coming from abroad and under the circumstances and uncertainties that prevail in Santiago de Chile indicates the “general trust” of decision-makers in academic work. It is, however, also possible that the decision to commit to the proposed activities was influenced by other factors associated with trust and a willingness to cooperate:

- good contacts between Chilean and German research institutes had already been established in the *Risk Habitat Megacity* research project, which ran for five years in Santiago de Chile (Heinrichs et al. 2012),
- contacts and ties to GORE had also been established within the scope of the *Risk Habitat Megacity* project,
- strong ties existed between some of the Chilean scientists and local stakeholders.

This links to Berrang-Ford et al. (2011, pp. 158) who pointed out the positive link between science and policies and its effective impact on decision-making under time constraints. “If a system is able to convert science into policies in a relative short-time frame, this system’s strategic planning will get supported by better promptness, responsiveness, and dynamic and flexible decision-making.”

Apart from producing the Regional Climate Change Adaptation Plan, the accompanying Manual for Implementation (Krellenberg et al. 2012a, b), and the above-mentioned briefing and working papers, scientists also introduced a Manual for Practitioners (Krellenberg et al. 2012c) and Information Material for selected municipalities (Krellenberg et al. 2012d). The Manual for Practitioners presents the overall findings on climate change and its impacts to a wider public, while the Information Material provides specific municipal results and recommendations for three selected municipalities. Again, all written material was in Spanish, using a non-scientific vocabulary. Print versions were disseminated in the MRS and documents made available on the project website as a free download. A Web-GIS—an interactive portal for processed and analysed data—was designed and made freely accessible. In the online questionnaire, stakeholders were asked for feedback on the information provided by the scientists involved in the project. Access to the minutes of roundtable meetings and the regular up-dates of ongoing initiative activities in Santiago were highly appreciated. The Information Material on the three selected municipalities was also perceived as a valuable asset of the overall product.

Although the scientists’ work was seen positively by the polled stakeholders, political leadership, including personal contacts and trust, was considered indispensable to its acceptance in practice. Due to the absence of an institutional or legal framework to which the Regional Climate Change Adaptation Plan could be ascribed, holding the interest of roundtable participants in the overall process and pushing the Plan forward to the final moment of handing it over to the politicians was only possible through the strong leadership of a number of individuals on both sides, scientific and practical alike. For the most part, however, it was the scientists who liaised between the participants and provided a well-organized frame that kept the process alive.

It is worth mentioning in this context that according to the online questionnaire almost all participating stakeholders would appreciate guidance and monitoring from international experts during the implementation phase of the Plan. This was endorsed by an official letter from the current Intendente (head of the Regional Government) asking the coordination team to consider future support.

12.4 Final Implementation as the Responsibility of Politicians

Once the Regional Climate Change Adaptation Plan and its 14 adaptation measures had been finalized, the process of adaptation was to a large extent affected by and dependent on political time frames and political will. This gave rise to a discussion on issues such as public consultancy, the improvement of existent measures and economic assessments. It was now a reality that the ultimate success of the overall process was no longer science-driven but ultimately in the hands of the local authorities responsible for implementation of the Plan and its measures. The scientists in question accepted this fact, adapted their time frames to the political agenda and modified the Plan as requested by the decision-makers. Efforts were also made to encourage Chilean experts to work on the economic assessment of the measures, a precondition for the legal endorsement of their implementation.

Taking into account that the scientific component of the process ended when the Regional Climate Change Adaptation Plan was handed over to the local authorities in Santiago de Chile and that the implementation process is still in progress, this section discusses suitable implementation mechanisms that build on the results presented in Chaps. 9 and 10.

Strong leadership will be crucial to the implementation phase of the Regional Climate Change Adaptation Plan if the topic is to be pushed to the top of the political agenda. Given the governance structures in Santiago de Chile, the sector divide was seen as a crux from the outset of the process. This presentiment became reality when the roundtable debate focused on responsibilities. It soon became apparent that the number and variety of actors and organizations involved in climate change adaptation constituted a major challenge in terms of implementing the selected measures. In brief, who does what and when became key questions to be decided. Asked in the online survey to name the actors relevant to implementation, respondents referred most frequently to the national government, the ministries concerned and the local population. SEREMIs, Association of Municipalities and civil society organizations were mentioned in second position. All of the respondents emphasized that the municipalities should be part of the implementation process at local level.

Referring to the constraints of implementing innovative adaptation measures, participants spoke of ill-defined roles and a lack of willingness on the part of political decision-makers. They stated an urgent need for continuation of the participatory roundtable process in the implementation phase and clarification of the specific roles of each institution. From a scientist perspective, it is therefore strongly recommended that decision-makers continue to focus on transparency, accountability, and cooperation among all stakeholders in the final stage of the process to ensure effective implementation of the measures, and ultimately a strong commitment to adaptive climate change response.

12.5 Concluding Remarks and Lessons Learned

As a final evaluation from a scientific perspective, the evolution of a Climate Change Adaptation Plan for the Metropolitan Region of Santiago de Chile in a German-Chilean science-policy initiative with a bottom-up approach can be seen as an innovation in Chile and will hopefully be implemented by the politicians concerned.

The strategy of involving different societal groups (decision-makers, civil society, affected private sectors and selected municipalities) from the outset helped to put the topic of climate change adaptation on the wider public agenda and to promote the implementation of adaptation measures. Instead of working with 'perfect' solutions, the relevant stakeholders were invited to jointly develop the measures in a participatory process. Local settings were taken into consideration by listening to the specific needs and interests of the stakeholders in question and incorporating them into the process. As a result of sound leadership, careful communication strategies and flexibility in the face of political agendas, the challenges—albeit manifold—proved to be surmountable.

Roundtable participants who replied to the online questionnaire confirmed that awareness of climate change and adapting to its consequences in the MRS was considerably enhanced by the participatory process. Furthermore, results show that participants accepted the leadership of the scientific team—a vital step on the way to creating a solid basis for trust and close collaboration. With the mandate for action orientation and the political commitment of the two principal decision-making partners in the MRS, the interest of the roundtable participants was maintained over a period of two and a half years. Respondents of the questionnaire confirmed that the local stakeholders involved in the process were those most relevant to the topic. Since the majority of roundtable participants were technicians without powers of decision, however, the political will to implement the Plan repeatedly stated by GORE and SEREMI MA protagonists at different stages of the process became a major signal for all of the actors concerned. The ultimate success of the process was to a great extent determined by a number of key figures. Participants indicated nonetheless that now more than ever strong administrative leadership was called for if the Plan was to be realized, the commitment of the various implementing organizations guaranteed and long-term decisions made.

Two serious constraints should not be withheld: time and human resources. Coordination of a process involving 25 scientists and approximately the same number of local practitioners and politicians over a two and a half year period, the organization of ten roundtable meetings, cultural and language constraints, and the task of keeping scientists and local stakeholders on board posed major challenges. Comprehensive communication and careful preparation of public outreach documents proved to be vital but also time-consuming.

A further constraint arose in the context of the economic evaluation of the measures proposed. One of the participating institutions at the roundtable meetings offered to take on this responsibility. The aim was to add economic calculations for each of the adaptation measures. The main obstacle was lack of data, which prevented quantitative modelling and called instead for careful estimates.

In the following, four major lessons learned from the inter- and transdisciplinary processes in Santiago de Chile are highlighted in an effort to facilitate adoption of the process by other cities:

1. *Collaborative work for a common knowledge base:* Although participants of the roundtable meetings came from different sectors, the creation of a common knowledge base was possible, as they were committed to working collaboratively and accepted the process design. Having the right people at the table allowed for serious discussion and the spreading of information through the relevant channels. By facilitating and extending participant exchanges in the frame of regular meetings, both contact and interest were maintained. Nonetheless, more regular meetings (within shorter time periods) might have made communication and the entire process easier for both participants and organizers.
2. *Effective communication strategies and work on selected case studies:* The use of dissemination and a variety of products and publications, such as briefing papers, manuals, information material and public events, was crucial to bringing the topic of climate change and adaptation response to the agenda and keeping it there. In-depth research for three selected municipalities within the MRS allowed for demonstration of expected problems, including detailed data and responses to concrete practical examples. This led to a better understanding of how action could work and have a positive effect.
3. *Legitimacy for implementation:* The participation of a wide range of actors in developing the measures in question, the leadership of the two key organizations responsible for climate change response activities in the MRS, and the scoping studies undertaken to assess the feasibility of the measures was key to reaching implementable action. Another important step towards legitimacy was the design of low-cost measures, proving that response to climate change does not necessarily call for big budgets.
4. *Stakeholder feedback for practical outcome:* Finally, evaluation of the process with an online questionnaire allowed for overall conclusions to be drawn, as expectations of both practitioners and politicians were taken into account. This procedure could contribute to facilitating adoption of the process in other world metropolitan regions.

Regarding the evolution of adaptation in the future, respondents saw the participatory process of the scientific project as a first step. It set a precedent, provided information and contributed greatly to bringing the topic to the public eye. The wish was expressed for similar initiatives to take place more often and in other regions. This knowledge transfer has already been requested by a number of cities involved in the Regional Learning Network (see Chap. 11).

It is hoped that this book will contribute to further exchange and mutual learning between cities all over the world. In this context it should be highlighted that numerous cities are dealing with challenges similar to those of the MRS and have, in some instances, been proactive (Hansjürgens and Heinrichs 2013; Krellenberg et al. 2013). Others are still waiting in the wings and might likewise benefit from the approach. This overall integrative inter- and transdisciplinary approach developed for Santiago de Chile can be applied to other local contexts. In which case, keeping networks such as the Regional Learning Network alive and organizing frequent meetings between different cities throughout the world to strengthen mutual learning could be crucial.

References

Berrang-Ford, L., Ford, J. D., & Paterson, J. A. (2011). Are we adapting to climate change? *Global Environmental Change, 21*(1), 25–33.

Friedmann, J., & Stöhr, W. (1966). *The uses of regional science: Policy making in Chile.* Papers of the Regional Science Association 18, 207-22. Revised in Friedmann, 1973b, 255-72.

Hansen, T. N., Agapitova, N., Holm-Nielsen, L., & Goga Vukmirovic, O. (2002). *The evolution of science & technology: Latin America and the Caribbean in comparative perspective, background paper for LCSHD 2002 regional study: Knowledge in Latin America and the Caribbean: Econsidering education, training and technology policies.* Washington, DC: World Bank.

Hansjürgens, B., & Heinrichs, D. (2013). Megacities and climate change: Early adapters, mainstream adapters and capacities. In F. Kraas, S. Aggarwal, M. Coy, & G. Mertins (Eds.), *Megacities: Our global urban future.* Heidelberg: Springer.

Heinrichs, D., Krellenberg, K., Hansjürgens, B., & Martínez, F. (Eds.). (2012). *Risk Habitat Megacity.* Berlin: Springer.

Holm-Nielsen, L., & Agapitova, N. (2002). *Chile – science, technology and innovation.* Washington, DC: World Bank.

Krellenberg, K., & Barth K. (2014). Planning collaborative climate change adaptation responses in the Metropolitan Region of Santiago de Chile. In Stauffacher, M., & Gross, M (Eds.), Interdisciplinary Science Reviews, Vol. 39, No. 4.

Krellenberg, K., Jordán, R., Rehner, J., et al. (2013). *Adaptation to climate change in megacities of Latin America. Regional Learning Network of the research project ClimateAdaptation-Santiago (CAS).* Santiago de Chile: CEPAL Serie Seminarios y Conferencias.

Krellenberg, K., Müller, A., Welz, J., Bräutigam, K.-R., Kopfmüller, J., Lehn, H., et al. (2012a). *Pedro Aguirre Cerda. CAS. Material informativo municipal.* Leipzig: Helmholtz Centre for Environmental Research – UFZ. 27 pp.

Krellenberg, K., Müller, A., Welz, J., Bräutigam, K.-R., Kopfmüller, J., Lehn, H., et al. (2012b). *ClimaAdapataciónSantiago – CAS. Manual para profesionales.* Leipzig: Helmholtz Centre for Environmental Research – UFZ. 35 pp.

Krellenberg, K., Müller, A., Welz, J., García Soler, N., Bräutigam, K.-R., Kopfmüller, J., et al. (2012c). *Plan de adaptación al cambio climático para la Región Metropolitana de Santiago de Chile.* Leipzig: Helmholtz Centre for Environmental Research – UFZ. 63 pp.

Krellenberg, K., Müller, A., Welz, J., García Soler, N., Bräutigam, K.-R., Kopfmüller, J., et al. (2012d). *Manual de Implementación para el Plan de Adaptación al Cambio Climático para la RMS.* Leipzig: Helmholtz Centre for Environmental Research – UFZ. 32 pp.

Latorre R. (2001). *Science in Latin America: There is a hope?* IUPS Newsletter. 2001 cited August, 2013. 1–7. Available from http://www.iups.org/nl3/latorre.pdf.

Urzua, R. (2000). Social science research in academia and policy-making: The case of Chile. In U. Teichler & J. Sadlak (Eds.), *Higher education research: Its relationship to policy and practice.* Oxford: Pergamon.

Printed by Printforce, the Netherlands